The Other End of the Microscope:

The Bacteria Tell Their Own Story

Antonie van Leeuwenhoek

The Other End of the Microscope:

The Bacteria Tell Their Own Story

A Fantasy by
Elmer W. Koneman, M.D.

Illustrated by
Bert Dodson

Edited by
Eleanor S. Tupper

ASM
PRESS

ASM Press
Washington, D.C.

Copyright © 2002   ASM Press
                        American Society for Microbiology
                        1752 N Street, N.W.
                        Washington, DC 20036-2904

Library of Congress Cataloging-in-Publication Data

Koneman, Elmer W., 1932–
    The other end of the microscope : the bacteria tell their
own story : a fantasy / by Elmer Koneman ; illustrated by Bert
Dodson ; edited by Eleanor S. Tupper.
        p.   cm.
    Includes bibliographical references and index.
    ISBN 1-55581-227-9
    1. Bacteria—Popular works.   2. Microbiology—Popular works.
    I. Title.
QR74.8 .K664 2002
579.3—dc21

                                                    2001056672

Address editorial correspondence to: ASM Press, 1752 N St., N.W., Washington, DC 20036-
2904, U.S.A.

Send orders to: ASM Press, P.O. Box 605, Herndon, VA 20172, U.S.A.
Phone: 800-546-2416; 703-661-1593
Fax: 703-661 1501
Email: books@asmusa.org
Online: www.asmpress.org

Contents

Preface

For several years the corner of my desk has been graced with the tiny replica of a van Leeuwenhoek microscope. Standing no more than three inches high, a fixed lens is positioned over the tip of a tapered, needle-like rod which can be lowered and raised by a delicate screw mechanism. I can only imagine how startled this lens grinder from Delft must have been when, almost exactly 300 years ago, he first observed under his newly invented microscope—probably even more primitive than my model—his "animalcules" in a tiny drop of canal water carefully placed on the tip of the tapered-needle stage. As an amateur mathematician, using an elaborate method of proportionate measurements of the size of one animalcule against the elements in a grain of sand, he calculated that each drop of water of that volume contained 27,000,000 of the tiny creatures.

Now modern-day microbiologists, using sophisticated compound microscopic lenses and elaborate stains, rather dispassionately observe thousands of bacterial cells in each microscopic field of view, calculating how large they are, what shape they assume, and in what position they arrange themselves, in quest of establishing a presumptive identification. Looking beyond this strictly scientific exercise, the philosopher or poet might muse on what lies behind these tiny creatures that assume so many forms and align in such a variety of patterns. In one smear preparation the round ones hang around in pairs, seemingly never separating from one another, while in another preparation they string out as beads in long chains. In other arenas they cluster in tetrads or loose bunches, or may elongate and haphazardly arrange in cross stick–like patterns like characters of some mysterious Asian language, so that one could imagine a scholar might even be able to translate their message. What forces guide these species-specific presentations?

More as philosopher than scientist, for many years I have gazed on these microbes, almost as many times as they number in van Leeuwenhoek's small drop of water. The fantasy has often come before my mind as to what these "monads" would communicate, if indeed they could. During such moments of free association, comparisons often crossed in my mind with the rabbits in Richard Adams' *Watership Down*. Similarly, I imagined stories of the prokaryotes, as told through the "mind," "vision," and "voice" of these beleaguered micro-animals as they confront the huge human machines—antiseptics, antibiotics, autoclaves—that methodically annihilate them in their "burrows."

As tiny as each bacterium is, often appearing under the microscope as little more than a mote, it has a structural and functional complexity equal to any other living cell. These microbes collectively have the capability of adapting to virtually any environment, surviving in temperatures from beyond boiling almost to absolute zero; they can derive energy, with or without the use of oxygen, from substances ranging from carbon-free base metals to complex organic compounds. Through the power of their sheer numbers, and the ability to replicate to as many as 2^{36} times their original number between the rising and setting of the sun, they can experiment with countless mutations which, given sufficient time, can produce progeny able to withstand virtually any natural or artificial environmental change that could eradicate any other form of life. And collectively, bacteria generate a vast array of products, encompassing the stench of a barnyard and the amiable aromas of premium cheeses, fermentation products enhancing and preserving the finest of foods, and the acids, enzymes, and other complex commercial byproducts that have simplified the lives of humans in innumerable ways.

But, as happens in all living kingdoms, the bad actors also are part of the scene. Pathogenic prokaryotes invade the bodies of humans and other living creatures, destroying everything in their path, or producing toxins that wreak devastation many bacterial miles from their colonial habitats. Through the centuries the human toll has been enormous, and it continues almost unabated today despite great advances in modern medicine. These transgressions are also part of the complex relationship between humans and microbes.

As the rabbits of *Watership Down* found a human to tell their story, so I conceived the fantasy of Prokaryote Pond, a place in imagination where a vast array of microbes convene to interchange their nucleic acid–derived wisdom and make known to humans the grievances that have been accumulating for centuries. Here they find a forum to compare notes on the amazing variety of habitats they can occupy; to discuss the equally wide and marvelous range of their bacterial cell types and structures; to commiserate on the ravages of human-wielded antibiotics and compare strategies for defense.

Too, they can be envisioned as feeling strong discontent over the capricious manner in which humans have assigned taxonomic names to their ancient clans and families. Even humans in physician clans complain of the way their research-oriented colleagues keep changing the names of microbes and their progeny. The prokaryotes call for humans to be more thoughtful when coining the nomenclature for newly discovered organisms. More emphasis could be placed on affixing names that reflect the actual and positive attributes of a species, rather than simply recognizing the location of discovery, or honoring some individual human microbe hunter, no matter how illustrious. In their resentment, the bacteria of Prokaryote Pond offer in rebuttal a renaming of *Homo sapiens* that reflects their view of humanity's actual nature, having lost confidence that we humans are truly "wise."

My motivation in this book is to bring the viewpoint of the bacteria, these specks of complex dust, to the awareness of humans. The tales I relate here have evolved over many months, leading me on journeys into the back alleys of many human discoveries and arenas of microbial activity. A marvelous literature exists extolling the wonders of the microbes, their diversity, mutability, ambulation, metabolism, and mechanisms of antibiotic resistance. These stories, told here in their own "voices," are intended both to tantalize the most experienced microbiologist and to provide the inquisitive with a better understanding of the hidden world of the microbes.

So I invite you the reader to look in on Prokaryote Pond, allegorically transformed into a large microscope stage, allowing you as visitor or participant to look down the lenses of imagination, to experience a bit of fantasy which, as applies to much of mythology, embodies more truth than fiction.

Elmer W. Koneman, M.D.
Breckenridge, Colorado
March 2002

Acknowledgments

The first hurdle in the acceptance of any work is someone to recognize its worth. For *The Other End of the Microscope*, this person was my colleague Davise H. Larone, Ph.D., an established author and illustrator in her own right, whose enthusiastic response gave me the initial encouragement to move on with the project. After acceptance of the book by ASM Press, James K. Todd, M.D. (Children's Hospital, Denver, Colo.), and Margaret M. Yungbluth, M.D. (St. Francis Hospital, Evanston, Ill.), reviewed it and provided invaluable counsel on organization and on key bacteria to include.

My thanks go to the staff of the Colorado Association for Medical Laboratory Education, Lynn Maedel, Lor Childs, and Ellen McClurkin, whose sage advice and encouragement facilitated the early stages of composition. I am grateful also to the several score pathologists, microbiologists, and medical technologists who through the years have shared their experience with and insights into the challenges presented by the prokaryotes. Finally, I hold in highest regard Ellie Tupper, whose dedicated and detailed editorial handiwork has resulted in accuracy of content and a magical flow of words, and Bert Dodson, whose well-devised sketches bring wit and clarity where words fall short.

Prologue

Imagine a microcosm—literally a microcosm—of the smallest of creatures on Earth. From at least the time of the Greeks, humans have considered themselves the center of the universe, but these tiny beings existed far earlier, before even the ancestors of the ancestors of humankind. These invisible creatures compose the domain Bacteria: the original and most basic of organisms on the planet, preceding by almost 2 billion years the vast array of complex eukaryotes that later came to populate the world. Also called prokaryotes, these microscopic one-celled beings are so ancient they lack even a nucleus. Nonetheless, they manifest a multitude of characters that only now, at this late date in evolution, are being recognized and understood.

Here at last these tiny creatures have a voice. The time has come for the bacteria to tell their story, a tale of wonder, mystery, and surprise. With their flagella waving, their colorful coats ablaze, their gases bubbling and odors surging, they command our attention.

A dream, you say? Perhaps. A fantasy? Maybe so. But listen and learn how it might be—how it is. Immerse yourself in Prokaryote Pond.

A Time of Wonder

... The King of Glory in his powerful Word
And Spirit coming to create new Worlds[:]
On heav'nly ground they stood, and from the shore
They viewed the vast immeasurable Abyss,
Outrageous as a Sea, dark, wasteful, wild,
Up from the bottom turn'd by furious winds
And surging waves, as Mountains to assault
Heav'n's highth, and with the Centre mix the Pole.
Silence, ye troubl'd waves, and thou Deep, peace ...

—John Milton, *Paradise Lost*, VII:208–215

In a remote and mysterious location lies a small pond of water, measuring no more than 10 meters at its broadest. The pond rests in the bottom of a shallow depression of land surrounded by low hills, much like a saucer or, perhaps more appropriately, the parabolic curve of a giant radio telescope.

Shortly after dawn a sultry, whirling wind rises from all directions—north, south, east, west, and all points in between. Never before, except perhaps at the time of the world's creation, has there been such a wild gale; it tears furiously across the surface of the water in the pond, whipping it up like a giant vortex mixer. From bottom to top the water is tossed and roiled, almost escaping but held in by the rim of land on all sides. Seaweed-like plants are torn to shreds, releasing abundant proteins into the water. Red algae, firmly rooted to large stones at the pond bottom, are dislodged, releasing water-soluble pigments so that the pond turns distinctly blood-red, like a puddle in a slaughterhouse. Secretions and extracts from myriad tiny animals, both dead and alive, add carbon and nitrogen, amino acids and traces of carbohydrates, disaccharides and monosaccharides, hexoses and pentoses, dozens of other organic molecules. Minerals boil up from sediments stratified for centuries beneath the superficial sludge at the

water's bottom—calcium, magnesium, manganese, iron, zinc, and other monovalent and divalent cations—balancing the anions that are leached by the violent swirling of the water from piles of stones lying along the shore.

The winds then stop as rapidly as they began, and the water falls suddenly calm. The warm water and the sunny morning air reach thermodynamic equilibrium, balancing at a favorable ambient temperature for bacterial growth. Proteins, sugars, minerals, all gel into place. The pond, in effect, becomes a giant agar plate, containing exactly the right concentrations of substances—organic compounds, balancing salts—for the survival of the greatest variety of bacteria.

Now a thin cloud of dust, blown up by the powerful winds, begins to settle onto the solidified surface of the pond. Adhering to the microscopic dust particles are spores and cell bodies from virtually every species of prokaryote that has ever been described, and more that humans have never before seen or recognized. Only in the account of Noah's ark has there been anything like this, an event where all of the species of a given kind in the world are gathered in one confined area.

There are organisms already naturally living in the pond: the resident prokaryotes—bacteria and some blue-green algae, or cyanobacteria—together with a vast assortment of microscopic eukaryotes, including fungi, slime molds, flagellates, green algae, and other intriguing creatures. But the bacteria are the ones who emerge first, seizing the opportunity to replicate and expand in this habitat ideal for them. Across the firm red surface of the congealing pond, one bacterial colony after another arises, all displaying their own unique characteristics.

Soon a kaleidoscope of colors and variations in colony types has appeared. The creamy yellow colonies of the staphylococci and micrococci come up early; interspersed are the smaller, transparent colonies of the streptococci, some surrounded by a clear, circular walkway where the water's red pigment has been dissolved away. In the distance a few brilliant, deep red colonies of *Serratia* lie interspersed with green, spreading patches of *Pseudomonas*, the pigment of which leaches into the surrounding agar gel. Here and there near the shoreline, looking like grains of pepper on a smear of ketchup, scatter the black colonies of *Salmonella*; hydrogen sulfide gas has escaped through their cell walls and reacted with ferric salts in the environment to produce a black complex.

Similar black colonies of *Proteus* can also be seen; fortunately, only the nonmotile members of this genus have found their way onto the surface of the pond. *Proteus mirabilis* and its cousins would have rapidly swarmed over the visible veneer, making it difficult for other bacteria to grow. Against the horizon a prominent skyline develops: the upward-growing, brilliant white colonies of *Streptomyces*, rounding like ant hills and permeating the air with their distinct earthy aroma. Also distinctive about *Streptomyces* is the absence of bacterial growth in its near vicinity, the effect of noxious substances that render its immediate colony perimeter uninhabitable, like the barrier moats of medieval castles, holding off all comers.

The array of emerging colonies soon surpasses any attempt at description. A few fungal spores in the water, along with some blown in with the wind, also show signs of germinating. They are a definite threat, as they could cover the pond, pushing the bacterial colonies to the outer reaches where they would soon dry out and perish. But because fungal spores require a more prolonged germination time, their growth is of no concern for the moment. Protista and other eukaryotes will soon also arise. But for now the bacteria alone rule this unique and miraculous microcosm.

Let us call it "Prokaryote Pond."

Thermotoga maritima

And so, let me introduce myself. I am *Thermotoga maritima*. Who is that, you ask? Who am I? Ah yes—though I am a prokaryote and a bacterium, none of my clan ever appear in your laboratory cultures. We are extremophiles; we live in a realm outside of your circle of diseases. This obscurity will perhaps have its advantage, as we are unbiased, uninvolved in the tales you are about to hear.

We live where almost no one else would ever want to—deep in the sea, in the environs of the submarine volcanoes called hydrothermal smokers, where the mineral-laden water reaches temperatures close to boiling. How do we survive there? Just one of the many mysteries you will learn in these tales. We in effect don a hot suit—an insulated protein outer mantle, which looks rather like the togas worn by the ancient Romans. So you can see how Dr. Karl Stetter arrived at our uniquely descriptive name: *Thermotoga*—the heat-loving, toga-wearing prokaryote that lives in the sea ("*maritima*").

I am a prokaryote, more specifically a bacterium, a member of the domain Bacteria. As you probably know, this means a single-celled organism with no nucleus or other membrane-bound organelles. The domain Archaea are also prokaryotes, but some of their characteristics are very different from the bacteria. Humans group all the rest of life on the planet into the domain Eukarya, which contains the protista, fungi, plants, and Animalia—including you humans. But I speak today only for my bacterial cousins, gathered here in a unique assembly to tell you their stories.

And so I welcome you to Prokaryote Pond. We hope here to show you humans a world you have rarely seen, possibly beyond your imagination. To our regret, often to our harm, human beings have historically considered microbes to be "germs," "bugs," the enemy—if they have considered us at all. Rarely have humans understood what microbes truly do and are, and the secret and various services we perform for the benefit of the entire globe. In fact, life as you know it would simply not exist, were it not for the bacteria. But no, we are reviled, "sanitized," misunderstood, misnamed. This is your chance now to listen to the voices of beings infinitely smaller, and yet infinitely more numerous and varied, than you: the tiniest creatures at the other end of the microscope.

These tales will show you just how diverse we prokaryotes really are, our infinite variety of habitats, and how we have managed to cope in each. Some human scientists believe that the earliest life started in the depths of the mud and muck of ponds and lakes, where certain of our anaerobic ancestors dwelled at a primeval time before what has been called the "oxygen holocaust." Some ancestral prokaryotes have even been found buried deep inside solid rock, showing that we can live almost anywhere. In the most uninhabitable of environments, in virtually any place in this universe where there is a source of energy, from freezing polar caps and Alpine lakes to the stinking, boiling sulfur pots of Yellowstone Park, you will find us. But I will leave these stories for my cousins to tell as this narrative unfolds.

It may gall some among my human audience to realize that we simple one-celled creatures belong to a kingdom created long before any of you. Oh, I know, biblical scholars among you will point out that we microbes were never mentioned in Genesis when Noah filled his ark "two by two." But, I assure you, we were there. Each of those pairs of animals carried a whole host of prokaryotes aboard, hidden from the wise old ancients. For example, if my good friend *Escherichia coli* had not been present in the bowels of those select pairs of animals, industriously producing the essential vitamin K, each animal would have bled to death even before it crossed the gangplank. It may be even more difficult for humans to accept that, through centuries of gene transfers, chromosomal recombinations, and mutations, we prokaryotes have been able to fashion, and pass on to our offspring, techniques for survival against the huge variety of defense mechanisms invented by the animals and humans with which we cohabit—sometimes even before those mechanisms were invented!

Another purpose for our gathering today is a form of protest. Many of my bacterial colleagues simply are not happy with their treatment by humans during the several decades since our discovery; indeed, we have a long list of bitter grievances to present. One truly sore point is the way some prokaryotes' names have been changed and demeaned through the years, often trivialized by accentuating defects in their characters. For example, the species name "*anitratus*" is often appended arbitrarily to the names of some of my colleagues who are not endowed with the enzyme system necessary to reduce nitrates. "*Paucimobilis*" is a species name used to designate some among us who are immobile or unable to move very fast. Prokaryote names are sometimes linked with those of humans who may be of questionable character, or with obscure places that are difficult to find on any world map.

In point of fact, some of my colleagues feel so strongly about these taxonomic slights, they have demanded a means for reciprocal action. Several of those clans who have had their names repeatedly revamped have formed a committee to discuss and present, in retaliation, a new name for *Homo sapiens*—one which will reflect our own perspective of the human race. This debate will occur in the final session of our assembly. It should be interesting to see what these offended members come up with.

Discontent has also run high among certain bacterial species over being worked to the point of exhaustion, as in pickling vats and industrial fermentations, only to be blamed for producing the alcohol and other compounds that have caused such ravages among the human race when misused. You also will hear of the calumny borne by others of our kinsmen who are referred to by humans as "a scourge" or as bacterial "rogues" for the misery and diseases they cause. In rebuttal, several of us will eloquently point out that most outbreaks and epidemics of bacterial infections result from the mistakes of *humans* blundering into certain habitats where the prokaryotes have reigned since time immemorial.

More dismaying still, some of us prokaryotes have recently gained additional notoriety by being selected by certain evil humans to inflict harm on other humans: what is called biological or "germ" warfare. Many among our ranks are incensed by this term. We resent being used as an instrument of conflict: this mode of aggression is not a germ war, it is a *human* war! We are held hostage, with certain of our Nature-given predispositions taken advantage of to cause human illness and panic. But who gets the blame? The headlines read, "Anthrax Threat Raises Nationwide Fears," as if it was our idea!

As will be repeated in the stories of several of our colleagues, the bacteria have no *innate* intent to do harm. Whether their actions are damaging or fatal, or just bad manners, they merely play out what is encoded in their chromosomes. It is true to every species and family of organism on Earth that bad actors can suddenly appear. DNA, the common thread of all living matter, often misfires as it divides and reassembles in subsequent generations. Even humans unintentionally produce malcontents, thieves, murderers, and the like! We bacteria are at a disadvantage because we produce progeny in a matter of minutes, and the rogues among us emerge more rapidly to draw public attention. But even they simply act as they are created to act, their sole purpose being to survive and to replicate. We ask you to listen as each storyteller explains its own unique situation.

Thus the agenda of this singular assembly lies before you: stories and protests, good news and bad. I have been asked by my colleagues to serve as narrator, to introduce and explain these proceedings to our human audience. I have accepted this challenge somewhat reluctantly, as many others of my colleagues are equally qualified to carry out this task. But as an extremophile and not closely related to the bacteria here assembled—it's said we *Thermotoga*s carry genes similar to both Bacteria and Archaea—it is reasonable to say I have a more objective point of view. Thus I am pleased to accept this duty, and propose we begin our narrative.

Visualize Prokaryote Pond now, if you will, as the stage of a huge microscope, with you the reader looking down through the long lenses of imagination to find a world of minuteness that is both intricate and mysterious. The bacteria at center stage will relate tales of wonder that evolved over billions of doubling times, explaining their unique place in the universe. Much like the clown Tonio in the oft-performed human opera *I Pagliacci*, I shall assume a position at stage center, standing before you to ring in a saga that we hope may give humans many thoughtful moments. *Si può? si può? Signore! Signori!*—"If I may, if I may, sirs . . ." *Io sono il Prologo*—"I am the prologue." Here on this microscopic stage, against the unique backdrop of Prokaryote Pond, you behold us, the bacteria, incarnate, taking our rightful place among nature's biota. *Andiam. Incominciate!*—"Come, let us begin!"

The Assembly Begins

Thermotoga maritima

Near the center of Prokaryote Pond, the tip of a craggy bit of rock projects above the surface of the water. Sharpen the focus of your imaginary microscope and you will observe, wedged within a narrow split in the rock, a tiny twig that will serve as the podium from which the orations of this assembly are to be presented. In fact, just at this moment, Escherichia coli, *elected as chair of the assembly (so honored as one of the most widely studied of the prokaryotes, most versatile in its extent of animal and human colonization, and most innovative in its varied expression of virulence mechanisms), ascends the twig podium . . .*

With flagella proudly waving in the slanted rays of the rising sun, *Escherichia coli* viewed the marvelous landscape surrounding it [being devoid of sex chromosomes, prokaryotes are neither male nor female], with the myriad bacterial clans scattered across the crimson surface of the pond. Each of the bacterial cells within the strikingly arrayed colonies was metabolizing in harmony, expectantly awaiting the proceedings. With some pomp, *Escherichia coli* called to order this first and probably only convention of these minute creatures.

"Fellow colleagues!" Its words wafted across Prokaryote Pond. "Pause in your reproductive doubling for just a few moments to hear what is coming in this assembly. We have an impressive agenda of focus sessions, in which interesting and important information concerning the life and times of us, the prokaryotes, will be presented. The scope of this assembly will be broad, but nonetheless the information to be presented is interesting and in some cases vital to our well-being. Therefore, I invite your indulgence and suggest that you slow your metabolic rates approaching stationary phase, better to ensure that the available nutrients will be sufficient for the span of this assembly. As it will be impossible to assimilate all the information presented, the key points in each presentation will be recorded in the DNA of transferable plasmids which will be widely conjugated among you for future reference."

The Other End of the Microscope

Escherichia coli then waxed eloquent for a few moments longer, giving a brief description of the amazing climatic events leading to this gathering and pointing out the potential significance of this remarkable meeting. In addition, *Escherichia coli* emphasized one of the key ground rules for this convention: only full names would be used when addressing all presenters and participants, in contrast to humans who rarely spell out bacterial genus names in their lectures and writings, lazily truncating the genus name to a single letter. In human circles, the name *Escherichia coli* is usually simply shortened to "*E.*" *coli*; here, the prokaryotes would at last enjoy the full respect of their peers.

"At this juncture," *Escherichia coli* added, "allow me to acknowledge our assistant, a transposon or 'jumping gene' called Genie Transposon. In nature, transposons are small defined regions of bacterial chromosomes or plasmids, that is, small segments of coded information that hop, or transpose, from one strand of DNA to another, or to plasmids by which they may then be transported piggy-back style from one bacterium to another. The genes carried in some transposons may be of no use to the recipient; in other cases, important new traits may be acquired, turning a placid bacterium into a raging pathogen, or even the reverse. Genie Transposon is here to help disseminate the information shared here at this historic assembly. Genie, please post today's agenda on the kiosk here by the podium."

With a sudden bound like a jack-in-the-box, the diminutive transposon leaped to the kiosk, a fragment of leaf wedged next to the podium twig, and affixed the list of planned sessions.

"Carefully take note of the agenda," *Escherichia coli* requested. "I remind you that prior to Session 4, each of you who wish to participate in the competition for the renaming of *Homo sapiens* must submit your entries to either of the copilots of the Taxonomy Committee, *Stenotrophomonas maltophilia* or *Serratia marcescens*."

Escherichia coli pushed a few of its flagella to the side so it could better view the assembly, then continued its introductory remarks. "I will next introduce to you our eminent colleagues,

First Congress of the Prokaryotes

Keynote Addresses: *Moraxella lacunata* and *Acinetobacter baumannii*

Session 1: Habitats and Niches
Coleaders: *Legionella pneumophila* and *Helicobacter pylori*

Session 2: Bacterial Structure and Function
Coleaders: *Micrococcus luteus* and *Enterobacter aerogenes*

Session 3: Microbial Pathogenesis and Human Infection
Moderator: *Stenotrophomonas maltophilia*
Panelists: *Staphylococcus aureus, Pseudomonas aeruginosa, Escherichia coli*

Session 4: Antimicrobial Mechanisms and Defenses
Coleaders: *Enterococcus faecium* and *Bacillus* JK

Session 5: Final General Assembly: the Renaming of *Homo sapiens*
Presiding: *Escherichia coli*

Moraxella lacunata and *Acinetobacter baumannii,* who will present the keynote addresses for this general assembly.

"Each of these speakers was selected from among our respected states-bugs as one whose dignity has been most offended by humans over the turbulent decades since we were discovered. Humans have had the audacity to enact name changes for each of these two honored colleagues at least a half-dozen times. In compensation for these insults, these honored members, along with other bacterial brethren whose names have been changed repeatedly at the hands of humans, have been asked to lead the *Homo sapiens* Taxonomy Committee. As you may know, sentiment is running high among many committee members that the species name *sapiens*, meaning 'possessing or expressing great sagacity and discernment,' can, from our perspective, no longer be applied to the human race. Hence, the sole charge of this committee is to deliberate and bring back to this assembly for final vote the several proposed name changes for *Homo sapiens*.

"The grand prize for the winning name entry," continued *Escherichia coli* with flagella flying, "will be the insertion into the winner's genome of a plasmid carrying all genetic information necessary to confer immunity, in perpetuity, against any form of attack from any host, be it animal or human. This offer is not just for the protection of the winner and its progeny, but in a spirit of mischief to promote the emergence of yet another resistant mutant to intrigue and frustrate human scientists.

"Now, honored colleagues," *Escherichia coli* concluded, "I introduce to you *Moraxella lacunata*, who in turn will set the stage for its cousin, *Acinetobacter baumannii*. Although stories of their shifting identities at the hands of human taxonomists are well known in some circles, others of you have not been fully briefed. Through their keynote comments, all will gain new and added perspectives. Without further ado, I present our distinguished colleague, *Moraxella lacunata*."

Moraxella lacunata, being incapable of self-motion (humans call this "nonmotile"), was "piggy-backed" by *Capnocytophaga* to the tip of the rock podium, allowing a clear efflux in all directions. (*Capnocytophaga* species fluidly crawl over solid surfaces without the benefit of flagella, a phenomenon known as "gliding motility." The movement is managed through the presence of special macromolecules protruding from the outer membrane of the cell, which successively expand to anchor to the substrate, contract, and relax, to move the cell forward like oarsmen gliding a boat over the surface of the water.) The assembled bacteria marveled as they observed *Capnocytophaga* smoothly ascending the rock surface, with *Moraxella* balanced high on its top side.

As the keynote address was about to begin, most of the microbes present, though respecting *Escherichia*'s counsel to

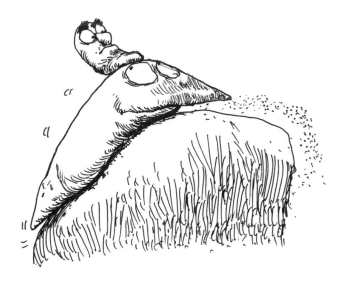

maintain a calm metabolic mood, nonetheless shifted into log phase of growth, both increasing the intensity with which they could listen to the emanations from the orators and also occasionally exuding a variety of metabolic products in the form of applause that wafted over the surface of Prokaryote Pond, giving a pleasant aroma to the surroundings.

Moraxella lacunata's Keynote Address

> In all falling rain, carried from gutters into water-butts, animalcules are to be found; . . . and that in all kinds of water, standing in the open air, animalcules can turn up. For these animalcules can be carried over by the wind, along with the dust floating in the air.
>
> —Antonie van Leeuwenhoek, 1702

"It is a great honor that I should be selected as copresenter of this keynote address. Certainly others more famous than myself could have been chosen—*Bacillus anthracis*, the first among us whom humans linked with a disease; *Mycobacterium tuberculosis*, from whom you will hear later on, the creator of the "White Plague" which for centuries has literally taken the breath away from the human race; *Yersinia pestis* (who also experienced name changes, from its initial designation of *Bacillus pestis*, to *Pasteurella pestis*, to its current tag after a Swiss-French human, Alexandre Yersin), our famous brother who was responsible for that horrible disease that humans call the Black Death; or *Vibrio cholerae*, an aquaphile that has reigned supreme for centuries through regional and worldwide pandemics.

"In reference to name changes, our clan *Moraxella* has faced disruptions on more than one occasion. Yet, we have had merely a fraction of the troubles of my look-alike cousin, *Acinetobacter anitratus* or *baumannii*, whose family has at several points in time suffered a distressing identity crisis. But I shall let my cousin tell you of this saga in its own words. Indeed, other prokaryotes have also experienced several name changes, and I hope some of these colleagues will rise at the conclusion of this address to offer, in return, their suggested alterations of *Homo sapiens.*

"Before addressing this business of name changes, however, I first would like to offer perspectives on the historical and scientific position that we, the proud prokaryotes, occupy in the world. I suppose, along with all other creatures, we are considered—at least by humans—to fall under their dominion. Although we bacteria are not specifically mentioned in creation stories, we indirectly fit in the Hebrew version, under the admonition of Yahweh shortly after creating the human race: 'Rule over the fish in the sea, the birds of heaven and every living thing that moves upon the earth.' Well, we were among those that 'move upon the earth,' although not all of us move so nimbly. When Moses recorded this bit of history, he obviously did not know about us, but we were present nonetheless, or he wouldn't have been alive to write it.

"(And if anyone wonders, at any point in these proceedings, at our command of human literature—let me remind them that as each human word ever written was set down, *we microbes were already there!*)

"For thousands of centuries," *Moraxella lacunata* continued, "we remained hidden in our marvelous microscopic world. We were able to live and conduct our business without human intervention, free from the interference and abuse we have suffered in these later times. At times this meant that human beings, and their animals and plants, were affected and even harmed. As a result, before they recognized who we were or that we even existed, humans walked the earth in fear of us. Some of our more hazardous brethren at times virtually had humanity on the ropes, for example *Yersinia pestis*—though, as it will explain, it was minding its own business in its narrow niche among rats and rat fleas. Humans simply got in the way of this centuries-old ecosystem.

"It's what humans call the 'survival of the fittest,' this attack of one creature on another. Lowly plankton and diatoms, passively floating around in nature's waterways, are gobbled up by proto-zoa and the like, which in turn are devoured by copepods and snails, which in turn are subdued by mollusks, crabs, and little fish, which are eaten by larger fish, which are eaten by polar bears, which, at the end of the line, are murdered by humans for their meat and other products. Humans dispassionately call this 'the food chain.'

"But then why should they become disturbed when we, the lowly prokaryotes, use our metabolic products in keeping with this survival-of-the-fittest business? It's not our intent to harm people, just to survive. In contrast, witness what some humans have done to one another. Oh yes, we have bad actors among our prokaryote clans. But as for humans? What about Genghis Khan and his Mongols, the Huns, the perpetrators of the Inquisition and the Soviet gulags, the Nazi exterminators, the international terrorists of modern times? Villains all, I would say, who collectively have eliminated huge numbers of their own kind. So is it their DNA? Is there something inherently evil in the millions of combinations of base pairs in certain chromosomes? Genetic aberrations coding for harmful behavior can occur in the simple cells of bacteria; do they also appear in the multifunctional cells that make up the human form? I cannot say. My point is that among the bacteria, as among humankind, the vast majority are not murderers.

"Therefore, friends, let us not indulge in too much self-recrimination over the way humans have suffered from many of our bacterial brethren. The Hebrews lament, 'The serpent was more crafty than any wild creature that the Lord God had made.' The serpent's nature is innate in the assembly of its nucleic acids—the snake in the garden is there—it is right there in the chromosomes.

"It was fated that we prokaryotes would eventually be discovered. The human mind is simply too advanced to remain forever in darkness. Yet we remain amazed, and also joyful, that our discovery required the span of so many centuries. Prokaryote-related diseases have plagued humans since the onset of time; yet, until only recently, unschooled minds blamed the conjunction of planets, the imbalance of humors, even divine wrath in retribution for misdeeds, as the underlying cause for the bane in which we have been secretly involved. Many of our clan-bugs perhaps became too bold in their assault on humans. Yet, even with attacks as severe as the Black Plague, our precision work was never suspected; rather, a widespread renewal of incantations, indulgences, intercessions, penance, and purgation was instituted to keep us at arm's length. With the results you would expect: ever hear of a microbe being expunged by a prayer?

"Yes, our discovery was largely fortuitous. The good Antonie van Leeuwenhoek, a 16th-century Dutch biologist who had no idea where his experiments with glass lenses would lead, became fascinated with how a tiny curved piece of glass could magnify bits of nature. He looked at the mouth parts of insects, described the circulating red blood cells, was the first to see living animal sperm cells, and discovered living rotifers, which he called 'animalcules,' in rainwater from a gutter. Then one day he focused on a smear of plaque from human teeth. There, to his astonishment, he found 'an unbelievable great company of living animalcules, a-swimming more nimbly than any I had ever seen up to this time. . . . in such enormous numbers, that all the water . . . seemed to be alive.' These were among the first observations of living bacteria ever recorded.

"Yet, even with this advanced intelligence, the better part of two centuries elapsed between van Leeuwenhoek's primitive sketches of his animalcules and Louis Pasteur's discovery of what marvelous powers we prokaryotes really possess. Then it was only a few years later that Robert Koch discovered how nasty we could really be if we got into animals and humans at the right places at the right times. Herr Dr. Koch really captured us, first in broth cultures, then on the surface of potato slices. There we were, stark naked, scrutinized under microscopes and hand lenses, probed from one end to the other.

"We have ever since endured the direct assault of an array of dyes, chemicals, and antibacterial drugs. We have been made slaves to serve humans in the production of countless substances. Yet, we fight back—we have not been annihilated. The story of this retaliation and antibiotic resistance movement will be retold later by the leaders among our resistant cousins.

"Let me return now to the main theme of my address: appellations, designations, cognomens, bynames, eponyms, labels, taxa. These really are human matters. Bacteria have had little interest in what they are called; that is, until so many name changes have been foisted on us that we no longer know who we are or what we represent. For example, my family has changed very little through the centuries. It is what has been discovered about us that has changed, an awakening that humans call scientific progress. During different decades, humans have found it necessary to puff themselves up by parading a new name among their colleagues, flaunting their superior intelligence at having made a unique new discovery. But for us, we are always what we are.

"As a clan, we the *Moraxella* have never been pretentious. We appear rod shaped (which humans refer to as 'bacillus' or 'bacillary'), and, as our cell bodies are barely longer than wide, the term coccobacillary is often used to describe us. We are very conservative in our growth, forming only small aggregates. And outside of a few metabolic excursions, we rarely cause anyone harm. In no way are we nomadic; in fact, one of our distinguishing characteristics is that we stay put—we do not move at all. We are without flagella or other means of motility. We are also very conservative in our eating habits. We avoid ingesting carbohydrates; we're inverse vegetarians, you might say, and depend instead on deriving our nourishment from preformed compounds. Some of our family are so lazy that growth supplements are needed when we find ourselves struggling on laboratory culture plates. The addition of a little serum or oleic acid to the culture medium for us is like feeding a fussy-eating human child a bit of ice cream.

"We might have remained unrecognized for many more years had it not been for two incidents, occurring almost simultaneously, when certain of our family members became a bit too bold and indiscreet. Just as certain enclaves of humans have found special habitats to their liking, we the moraxellae find the human eye—the conjunctiva, in specific—a virtual paradise. By taking up housekeeping in the eyes of only a few humans, widely isolated from one another, our role as the agent of what humans call 'pinkeye' (a rather pithy moniker) remained unnoticed, or unconnected. But now let me introduce Dr. Victor Morax, a Swiss eye doctor. At the turn of the 19th century, he examined several cases of conjunctivitis at Dr. Parinaud's clinic in Paris, during a period of study under Dr. Emile Roux at the Pasteur Institute. There we were—in the middle of the lachrymal secretions of patient after patient. We had nowhere to hide; we were caught in the act. He wrote down a detailed description of how we looked and how we acted, and forever afterwards doctors have been alert to us.

"Now to our second exposure. On the scene came Dr. Theodor Axenfeld, a private practitioner at the university Augenklinik in Breslau, a wonderful city nestled in the valley of the river Oder in southwestern Poland. (Even though the city was officially known as Wroclaw to the Poles, who really owned the place, it was called Breslau because so many Germans lived there.) You can picture him scratching his head trying to solve the mystery of what was causing pussy eye infections in 40 of his clinic patients. But by then the word had been broadcast: one need only study a stained smear under the microscope. Again, there we were in his microscopic field of view, and there was no doubt in his mind that we were the culprits. He never saw an infection in any patient in our absence. Soon afterwards, our humble, reserved family found its name in the headlines of the world medical journals as the perpetrator of pinkeye—the 'Morax-Axenfeld' bacillus.

"This notoriety soon reached across the English Channel. Dr. J. W. Eyre, an ophthalmic surgeon working at St. Mary's Hospital for Children and Guys Hospital in London, knew all about us. In his laboratory cultures he proudly recovered the 'Morax-Axenfeld bacillus' from the inflamed eyes of several patients. Then came another step taken in scientific progress. One of his laboratory assistants observed the presence of a peculiar depression or pit under our settlements (colonies), right where we grew on the surface of their culture medium. This good doctor connected our ability to 'eat out' the protein in the culture medium with the same property of 'eating out the eyes' which, of course, led to the eye inflammation in his patients. This was one of the first discoveries of how a particular bacterial property is linked to a specific disease process. Discarding our newly acquired names from the continent, he commemorated our pitting prowess with the name *Bacillus lacunata*.

"Then a whole sequence of new names followed one another. It may have been the poor resolution of the microscopes, or the distorted vision of the laboratory assistants of the day, but someone described us as spheres, always living together in pairs, and dubbed us *Diplococcus liquefaciens*! But about the same time, a fellow by the name of McNab, working in Germany, did not like the idea that these English blokes were robbing my family name of its association with the honored Drs. Morax and Axenfeld. In rebuttal, he tagged us *Diplobacillus moraxaxenfeldii* (a hard name to pronounce, but recognizing us both as a bacillus and with a connection to Morax and Axenfeld, yet still harboring the same mistake of thinking we exist in pairs).

"Finally, four decades following this flurry of activity, Dr. André Lwoff, while in service in the physiology-microbiology branch of the Institut Pasteur in Paris, purely as a side issue while engaged in study of the family *Haemophilus*, gave us our final family name

Moraxella, again recognizing the early work of Dr. Morax. It is a pity that Dr. Axenfeld's contribution to our family heritage somehow got dropped along the way, but this is the nature of humans. Obviously, as Dr. Morax was an alumnus of the Institut Pasteur, Dr. Lwoff was undoubtedly under some pressure to recognize his emeritus colleague. Remember also that Dr. Axenfeld was a private practitioner in a foreign land. More times than not, humans will recognize more the accomplishments of lofty university scientists, rather than those of the common grunts with hands-on experience in the clinical trenches.

"Following Dr. Lwoff's revised designation for our family name, many species names have appeared over the years, recognizing the aberrant behavior of certain of our progeny and next of kin. Humans are adept at noticing slight differences in behavior among us bacteria, and they seem to delight in cashing in on yet another name change. Thus, I should like to introduce my close cousins— *Moraxella nonliquefaciens*, a bunch who lost all interest in dissolving gelatin; *Moraxella phenylpyruvica*, another cluster who took great delight in turning phenylalanine into phenylpyruvic acid (a very narrow obsession in my view); and then, of all things, two cousins named after the cities in which someone first discovered them, *Moraxella osloensis* and *Moraxella atlantae*. These cousins live so far away that I really know very little about them.

"I shall bore you no longer with this purely human amusement. At least this gives you some idea how we must navigate through time among so many human foibles. Now I turn the podium over to my dear and nearest relative, *Acinetobacter*. It can give you even more of an earful over the number of times its family has had to adopt a new name. However, before yielding the podium, I would like to offer my suggested alteration of the name *Homo sapiens*. Based on the story I have just related, and the capriciousness with which humans seem to treat our noble family groups, I offer in nomination to the Taxonomy Committee the name *Homo desultus*, the surname taken from the French *desultus*, which means 'to leap down' or, more appropriately, 'marked by lack of a definite plan.' Thank you for your attention."

Spontaneous Effusions

A variety of vapors emanated from the aggregated colonies, as the form of applause given by bacteria in appreciation for an insightful presentation so succinctly given.

"Thank you, *Moraxella*," said the moderator, *Escherichia coli*, scanning the assembly to determine the level of interest. Noticing a sequence of bubbles emanating from slightly beneath the surface of the pond, *Escherichia coli* recognized *Clostridium tertium*, which wanted to share a particular insight. (*Clostridium tertium*, although possessing all of the metabolic machinery of an anaerobic bacterium, yet is what humans call "aerotolerant." An "obligate" anaerobe cannot live in the presence of oxygen, succumbing in only a few minutes of exposure. The aerotolerants, however, can survive in the open air, at least when growing on the surface of agar plates. Therefore, after coming to surface from under the oxygen-deprived water, *Clostridium tertium* could survive sufficiently long to address the assembly.)

Clostridium tertium

"Fellow prokaryotes," fermented *Clostridium tertium*, "I would like to share a development that is interesting to us, the anaerobic bacilli, and may as well be of considerable significance to the evolutionary history of prokaryotes in general. It is not our intent to enter into the strictly human philosophical argument of evolution versus creation, where the evolutionists speak in terms of millions of years of time between major advances, in contrast to the seven days of creation propounded by others. Time has no point of reference for us bacteria. We carry no watches, have no calendars, and only recently have been made aware of the temporal worlds of doubling times, 'lag,' 'log,' and 'stationary' phases of growth, and so on, all strictly of human invention.

"In any case, sometime within the span of evolution (however it is measured!), cells with nuclei, which are nothing more than compact aggregates of DNA wrapped in a protein-rich membrane, slowly developed. One theory postulates that nuclei arose from progressive incorporation of DNA from prokaryotes which lived together in symbiosis. Theory has it that certain bacterial cells merged with one cell line over a period of millions of years, accumulating more and more 'excess' DNA until a distinct nucleus emerged. The eukaryotes thus were born!

"Once eukaryotic cells were established, in the next billion years of evolution, complex colonies of cells evolved that ultimately became organs in 'higher' creatures, which in turn ultimately differentiated into animals, including humans. Within this theory is an astounding observation. Eukaryotic cells are respiratory, which means they must incorporate oxygen to extract maximal energy from food substances to feed their scheme of metabolism. Discovered within the extranuclear space—that is, within the cytoplasm of eukaryotic cells—are dark, membrane-bound, energy-rich organelles called mitochondria (*chondrion* = 'grain').

These actually can be regarded as tiny furnaces or micro power plants, fueling the energy for cell metabolism. Now here is the amazing hypothesis: mitochondria are actually thought to be an ancient breed of free-living bacteria. In the process of cell mergers and nuclear development, these primeval microbes were incorporated by eukaryotic cells as their personal oxygen-gulping, energy-manufacturing factories. This theory has

IN EARLY MICROBE HISTORY...

ONE MILLION (OR SO) YEARS LATER...

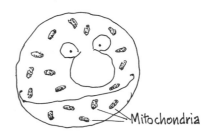

been championed by the human Dr. Lynn Margulis, whose impressive evidence to support it includes the observations that mitochondria not only have DNA that is distinct and unique from the nuclear DNA of the eukaryotic nuclei, but this DNA also has the capability of independent replication!

"Fellow prokaryotes," continued *Clostridium tertium*, "do you realize the implications if Dr. Margulis's theory is tenable? Many of our kindred serve as symbionts with humans, that is, happily living for mutual benefit on their skin, in their respiratory tract, and in their gut. They provide us protection, and we in turn, as our part of the bargain, secrete vital substances such as vitamin K, needed for blood stability. But beyond all that, we may in reality be an integral evolutionary part of every cell in every organ as well. Should this be true, not only is the idea *valid* that we preceded humans by eons of time—as we bacteria have held all along—but it is now recognized that our acquired intracellular presence is and has been integral and necessary for the function of any form of higher life composed of eukaryotic cells. Awesome!"

No sooner had *Clostridium tertium* submerged beneath the pond surface than *Leptospira biflexa*, with its characteristic corkscrew motion, wiggled upward to contribute yet another potential prokaryote-human association.

"What an amazing discourse, *Clostridium tertium*," undulated *Leptospira* with considerable enthusiasm. To the audience it continued, "I believe that most of you present know that we spirochetes have made our mark by our unique undulating and corkscrew motion. Many of the prokaryotes possess extracellular whiplike flagella, each anchored by a high-energy aggregate of proteins that serve as a 'pump' or 'nerve center.' The internalization of this motor pump and flagellum into our periplasm is the only difference we, the spirochetes, have from any of you. Thus, as this flagellum waves, our whole body undulates, much like the hip motion of a hula dancer. Obviously, possession of such flagella, either internal or external, provides the advantage that we can explore many environments and escape most immediate threats.

"I would like to add another wonder, in keeping with *Clostridium tertium*'s theory that mitochondria are independent, bacteria-derived organelles within eukaryotic cells. We have become aware of another theory, which proposes that the cilia possessed by certain cells, such as those lining the respiratory tree and those internal to the genital organs of animals and humans, *had their origin in an ancient strain of spirochetes!* In fact, this merged-spirochete theory has been extended to pertain also to the undulating organs of various other types of cells, including those possessed by protozoan ciliates and even the 'tail' of spermatozoa that is necessary for their self-propelled quest to fertilize an egg. Thus, humans may be a bit chagrined to learn that we prokaryotes, considered by many to be an enemy, in fact are integral to what makes their own inner systems tick. As the noted human scientist René Dubos once said:

> Microbes exhibit profound resemblances to man. They resemble him in their physical makeup, in their chemical composition and activities, and in their responses to various stimuli. They also display associations among themselves and with other living things which suggest illuminating analogies with human societies."

"Thank you, *Leptospira biflexa*, for this brilliant piece of insight," exclaimed *Escherichia coli*. "This new revelation of our extended importance in the world will have an impact on all future generations of prokaryotes.

"Realizing that this session is becoming a bit long, I nonetheless request your attention for a few more minutes before we take a pause. Getting back to our main theme, please listen carefully to what our next keynote speaker has to say. Indeed, its family name has also been bandied around through the years, and the upcoming narration promises to be cogent and to the point. I introduce to you our friend and colleague, *Acinetobacter baumannii*."

Just as *Acinetobacter baumannii* was about to proceed, a sudden darkness enveloped the area of the podium, like a solar eclipse, or a thumb over a microscope's light source. A sudden draft pushed through the dimness, as if a helicopter were about to land. Quickly the situation became clear. A *Xenopsylla cheopis* flea had unceremoniously landed atop the tip of the rocky crag, immediately next to the twig podium. In short order, reminiscent of Jonas's experience in the belly of the whale, an eructation occurred from the snout of the flea, and there in a fleck of clotted blood emerged the short, stocky forms of *Yersinia pestis.*

"My apologies for such a sudden and unconventional entrance," wafted *Yersinia pestis*, as the flea, with a mighty jump, vacated the scene, and the podium again became fully illuminated. "But reverberations in my chromosome indicated that references had been made in this assembly to certain activities of our clan. I appreciate your citation, *Moraxella lacunata*, and feel I should share a few additional aspects of our existence that should be of interest to those here assembled. Shall I continue now, *Escherichia coli*, or, as I am not part of the scheduled agenda, should my short discourse be delayed until later?"

"Well," returned *Escherichia coli*, "your mode of appearance, although unconventional, is nevertheless unique and impressive. It is evident that most of us here assembled are intrigued by what you may wish to share. A short interruption will not be out of place. Please proceed."

"Thank you. Indeed, it is our nature to perform our tasks in short order. As many of you know, the disease with which we are associated in humans and animals commonly has a rapid course.

"Whether I am here before you as hero or villain depends on which side of the fray you are on. Through the centuries my ancestors have been brutal, and we bear a legacy similar to that inherited by the perpetrators of other holocausts. Oh yes, we were an integral part of the many plagues and pestilences that afflicted ancient Rome, Egypt, Constantinople, Athens, and intermittently habitations all across the Gobi Desert and other points east. And, of course, in more modern times you have witnessed the havoc of the Black Plagues of the 14th and 15th centuries, in which tens of millions of humans worldwide were afflicted, not to mention an even greater number of deaths among rats and fleas. I do not excuse any of this, nor ask for justification, but a few bits of information might bring to light a somewhat different perspective.

"Bacteria have no ability to function teleologically. None of the plagues of history has been, at least from our part, directed toward any perceived end or shaped toward any final purpose, except to play out what has evolved in our chromosomes over millions of generations. Who and what we are is simply the result of a selection process among infinite possibilities presented in the mutations of centuries of chromosomal recombinations and the acquisition of plasmids along the way. No, we do not seek exoneration from our role in killing millions of humans by professing any doctrine of predestination, but—well, hear our story out.

"Our natural habitat actually is in the soil, particularly deep in burrows where remnants of dead rodents or insects may lie. In this state of existence, our ancestors were quite tame and harmless. But when this habitat became disrupted by wild rodents, usually rats, that dug new homes in these contaminated burrows, we gained access to the rats' tissues and bloodstream. This turned out to be a good relocation for us, even though once in a while a more susceptible animal could not withstand our presence and died. When fleas came by to dine on our host rodents, we came along with the meal, flushed into the gut of the flea along with the blood it ingested. So there we were, the three of us, rat, flea, and bug, coexisting in some degree of harmony.

"Then the situation began to change. It is not totally clear to us what happened, but certain of our ancestors assimilated key virulence factors. More and more infected rats began to die. Now, let me explain here that quirk in nature that may be fundamental to the evolution of the Black Death among humans.

"When the temperature of our environment is lowered, we yersiniae not only reproduce more rapidly, but also reactivate and increase the production of various enzymes and proteins. Picture us now, if you will, contentedly residing in the gut of a flea which has made its home on a nice warm-blooded rat. The rat dies, and the flea's ambient temperature begins to drop. Inside the flea's gut, as it cools, we begin to multiply rapidly and produce an enzyme known as coagulase, which acts to clot plasma. The end result is that the blood acquired by the flea during its latest meal is coagulated and forged into a bloody ball. This coagulum, supplemented by our rapidly reproducing microbial body mass, reaches a size sufficient to block the intestinal tract of the flea. Now the flea becomes extremely hungry and agitated and begins to jump aimlessly in search of a new blood meal. You witnessed this phenomenon here just a moment ago as my insect taxi, *Xenopsylla cheopis*, so quickly left this scene after coughing us out on this rock. (By the way, *Xenopsylla cheopis* is the name of the oriental rat flea, derived from *xeno*, meaning 'foreigner' or 'stranger,' and *psylla*, the Greek for 'flea.' Cheopis was a region in Egypt where this 'stranger' was first discovered.)

"Now to continue the story. These irritable, agitated, hungry fleas hop from rat to rat in search of a blood meal. If all the rats have been killed by the plague, these starving fleas can become absolutely ferocious. Enter an unsuspecting human; any warm-blooded creature that comes within hopping distance of the flea is fair target. The hungry flea's bite is deep and bloody, with literally hundreds of us being injected summarily into the skin of the human host. Once in the bloodstream, we are promptly engulfed by the white cells—but this has little effect as, also by a quirk of nature, we produce an envelope protein around our bodies that resists destruction by these scavenger cells. In fact, we use these monocytes as a second means of portage, taking up new residence in the first lymph node way station we come across. These swollen nodes, often the size of hen's eggs, are called 'bubos,' derived from the Greek word *boubon*, meaning 'groin' (because most flea bites are on the lower extremities of humans, the inguinal lymph nodes in the groin are the first to be affected). Humans call this form of the disease 'bubonic plague.' We also produce a substance called pesticin that triggers bleeding from small vessels throughout the body of any human we infect. These hemorrhages are seen as vesicles, eschars, and spreading pur-puric lesions that in time turn into large patches of black gan-grene. Thus the epithet 'Black Death,' and also our name exten-sion, *pestis*, from the Latin for 'plague.' "

The mention of the Black Death caused the emission of vapors from many colonies in the assembly. Taking this as a sign of inter-est and encouragement, *Yersinia pestis* concluded its tale.

"Humans really are their own worst enemies. In any situation where rats and their fleas are in abundance, pestilence will reign. At the time of the early Renaissance humans began to aggregate in close quarters in cities and hamlets. Sanitation was virtually nonexistent. Garbage and human and animal wastes literally clogged the streets. Meanwhile, trade routes were opened up between Europe and endemic plague regions in the Middle East by fur and spice traders who traveled both overland and by sea. The rats came with them, with fleas infected with us yersiniae along for the ride.

"The rats and their fleas invaded Venice, Florence, and Mar-seilles. Humans tried to stop us in Venice with a 'quarantine'—a decree from the Council of Venice that ships must be held in har-bor for 40 days (*quaranta giorni*) before unloading freight or pas-sengers. (One school of thought has the 40 days match the length of time Christ is said to have suffered in the wilderness; another that, after this length of time, it was found intuitively that the sur-viving cargo and humans could be granted safe passage.) We prokaryotes, of course, could survive all of this, and the plague continued.

"We wiped out the temporary Papacy in Avignon, where 90% of the inhabitants succumbed to our ravages. From there it was up the Rhine and on to London, with humans dying all along the way. Corpses in some habitations were said to be piled every-where that had space to receive them—in wells, caves, pits, pools, lakes, and rivers. The Christians blamed the Muslims, the Muslims blamed the Christians, and both blamed the Jews. Unknown to all of them, we were the cause of all this finger point-ing and holy wars.

"Thus, in closing, could it be that such a minute change in nature, a small readjustment of a tiny segment in our chromosome, could result in such immense human suffering? Could it actually be that a simple gene coding for the production of coagulase lies at the bottom of this scourge? Such a cause and effect is much too simplistic, and does not take into account the additional factors and forces I have related. However, this remains an intriguing speculation."

"Thank you," replied *Escherichia coli*. "What a story! You are welcome to stay for the remainder of the proceedings; and, indeed, are also invited to enter a name in the contest to rename *Homo sapiens*.

"Without further delay, then, my friends, please receive *Acinetobacter* with a warm welcome."

Acinetobacter took its place on the twig podium, and after scanning the crowd all around, nodded in recognition to all those assembled and began its discourse.

Acinetobacter's Keynote Address

> Minerals are classified primarily on the basis of their chemical constitution and secondarily on their crystal structure; and living things should be classified in the same manner, on the basis of their internal development rather than their external appearance.
>
> —Sigurd Orla-Jensen, 1909

"Fellow bacteria! I greet you on behalf of all members of our honorable family. Indeed, our grievances against *Homo* are broad and deep. There is a gentleness about *Moraxella* that permeates its soft words, the nonjudgmental nature of its discourse, its memory of the clips of history. My tongue will be sharper, my criticism more direct. I am not alone. In fact, our attack against *Homo* is more than just words, to which anyone working in modern-day hospitals can attest. Certain of our clan have become decidedly contemptible and have caused serious infections among many of the infirm. Humans call these infections 'nosocomial'; that is, those diseases originating or acquired in a hospital (the Greek *nosokomeion*, or 'hospital').

"But this is only a front. Are humans so out of touch as to believe that these infections occur mysteriously from the fault of the hospital? Not so at all. Human caregivers alone are to blame. They often fail to wash their hands after attending patients, they poke holes and insert tubes into every available orifice, and they compromise the immune system with powerful drugs and antibiotics. We bacteria cannot be blamed for taking advantage of their miscues, any more than a goat can be blamed for invading a garden if a fence has been broken down. We all have a right to survival. If humans cannot protect themselves, we cannot be blamed for setting up housekeeping wherever we can find a suitable niche. This is the same driving force that led humans into the darkest recesses of Africa and motivated passage across great seas to distant lands to set up colonies of their own.

"What is our chief grievance? The capricious name changes by *Homo* scientists that my family has had to endure. We support the new species name for *Homo*, '*desultus*,' as suggested by *Moraxella* (although we have an alternative that I will present later). We had no voice in the selection of our present name *Acinetobacter*, a tag less than honorable. Humans have the peculiar tendency to name bacteria after things they cannot do or tasks they cannot perform. '*Acinetobacter*' is from the Greek *akinos*, meaning 'unable to move.' Thus, we are the 'motionless bacteria.' What do they expect? We do not have flagella, cilia, or any other organs that are needed for motion. Why do they consider this a fault so egregious as to affect a family name? Should we rename humans '*aflugellatus*' just because they cannot fly?

"The insult does not stop there. One of our clan carries a second name also based on a faulty character. '*Anitratus*' is the species designation for any member that lacks the capability to reduce nitrates to nitrites. This name intimates that we have little talent: *Acinetobacter anitratus*, the bacterium that can neither move nor reduce nitrates. One talent that we *Acinetobacter anitratus* do possess is the avid fermentation of lactose, even in concentrations as high as 10%, with the formation of much acid. Why was this trait not honored with a name such as '*lactosus*' or '*lactosefermentans*'?

"Indeed, we did have one human friend, who actually was among the first to recognize us just after the turn of the century. Dr. Martinus Beijerinck was his name, and he dubbed us *Micrococcus calcoaceticus*, the latter because we have the unique ability to extract carbon from a simple medium containing only calcium acetate. Actually, he wanted to give us a German name, '*Essigbacterien*,' because we can also extract carbon from acetic acid, or vinegar (*Essig* = 'vinegar'). This indeed would have conferred on us a positive, honorable name. But the designation '*Micrococcus*' would not work: Dr. Beijerinck mistook our shapes for spherules, whereas we really are short rods.

"A pair of his lab mates, Isabelle Schaub and Frank Hauber, were on the right track after the second great World War by at least calling us 'bacterium.' *Bacterium calcoaceticus* would have been a terrific name. *Bacterium* or *Bacillus lactosefermentans* would have even been much closer to our nature and truly honorable as well. But they called us *Bacterium anitratum* ('It seems appropriate to select a species name that will express one characteristic,' is what Schaub said). For a short time we withstood the embarrassment of being called *Achromobacter* because our colonies were devoid of any pigment. Fortunately, a more intelligent scientist later decreed that the term 'achromobacter' is without taxonomic integrity, a courageous act that eliminated it from any official nomenclature.

"Perhaps worst of all was the heinous Dr. Jules De Bord, who threatened to divide us into two major clans and affix derogatory names to each. Those of us who have no interest in producing acids from carbohydrates were to be called *Mima polymorpha*. Do you know how this name came about? 'Polymorpha,' or 'multiformed,' was coined to describe our appearance when we are placed in certain stressful environments, particularly in the presence of antibiotics. When under stress, we stretch out, ball up, or become otherwise distorted in agony over the treatment we are undergoing. But we couldn't help it. Did you ever see the contortions a human goes through when it has been poisoned?

"Just as degrading, the name 'mima' was derived because, when humans observed us under the microscope in our natural state, they could not distinguish us from our near cousin *Neisseria*. This caused quite a stir in some circles when we were misidentified in genital secretions—in older times, women could have been socially ostracized for an indiscretion they did not commit! We are who we are—we should not have to bear a name that suggests that we 'mimic' or look like someone else!

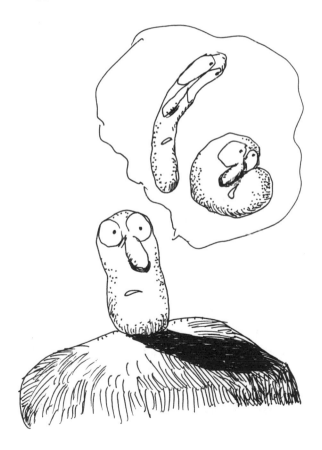

Fortunately, right-thinking humans have now discarded this ridiculous name.

"Even more embarrassing, Dr. De Bord decided that those among our clan who do have the capability to produce acids from carbohydrates were to be called *Herellea vaginicola*. When we inhabit so many ecosystems in the world, why should we bear the name that implies that we live only in a select part of the human female anatomy? Perhaps we should suggest that all gynecologists be called *Homo sapiens* subspecies *vaginiexplorans*!

"As for *Herellea*! What was that all about? Well, Felix d'Herelle, Canadian born, educated in France, while working at the Pasteur Institute made a discovery of utmost importance to us prokaryotes. He stumbled on an invisible microbe that was endowed with antagonistic action against the 'Shiga bacillus.' In fact, our *Shigella* colleagues did not know what had hit them. Their cells, living in the human intestinal juice, suddenly became metabolically deranged and mysteriously disappeared entirely from the dysentery efflux that Dr. d'Herelle was studying. This discovery occurred around the time of the first great global war, when more soldiers were succumbing to *Shigella* dysentery than to the ravages of gunpowder. Can you imagine the elation in the trenches that the Shiga bacillus was in trouble? For our part, we bacteria were undergoing ribosomal spasms (the only way we can perceive fear).

"In time, Dr. d'Herelle put the story together. You see, his 'invisible microbe' in fact was a virus, a bacteria-invading virus which he called a bacteriophage; that is, an agent capable of 'eating' bacteria. Of course, they do not really eat us, but rather penetrate into our inner substance and take over key ribosomal functions that affect protein production. In the case of the Shiga phage, a cadre of phage-infected bacteria produced 'nonsense' proteins which misdirected virtually all of their vital metabolic functions to the point of annihilation. But human elation was short-lived. We soon signed a mutual toleration pact with these phages, of much mutual benefit. In fact, many phages have become trusted colleagues, inspiring us to new metabolic accomplishments which, alas, sometimes produce even more human misery. For example, look what happened to *Corynebacterium diphtheriae*. Once it made the acquaintance of a collaborating beta phage, our gentle cousin was transformed into a toxin-producing terror!

"In the manner of humans, the right to honor Dr. d'Herelle for his discovery is fitting and proper; but why should he have his name applied to our clan, with which he had very little direct contact? This is nothing more than a tug of war between the French and the Germans, if you ask me. It would have made much more sense to connect his name with the Shiga bacillus, of which he did have first-hand knowledge. Indeed, *Shigella herelleae* would have been far more appropriate, but would have been vetoed as being the wrong genus and causing an unacceptable clash of ethnic cultures. But such a suggestion on our part is no more absurd than many of the bacterial names foisted on us by humans.

"Happily, subsequent taxonomists saw through all this and decreed that *Herellea* too was an illegitimate taxon. However, subsequent name changes did not necessarily improve our plight. The name *Acinetobacter lwoffii* came next, the species name honoring the Nobel Prize–winning French professor Dr. André Lwoff, also employed at the Pasteur Institute. Most recently, it's been *Acinetobacter baumannii*, named after a California microbiologist, P. Baumann, who has done the most exhaustive contemporary human study on our clan.

"Even so, what right do humans have to append their names to our honored, primeval selves? As so eloquently related by *Clostridium tertium*, we prokaryotes were around long before humans. Bacterial names are strictly human inventions. It is to the humans' benefit to tell us apart, not to ours. We bacteria have no inherent need for a name at all. Most aggravating is that humans, so often dissatisfied with a name, seem unable to make a permanent decision. Judge for yourself. Genie Transposon will put on the podium kiosk the following list of the names given to our noble clan at one time or another.

"Frankly, of all of these we would prefer the combination *Acinetobacter calcoaceticus*. We admit that we are not lively sorts,

Micrococcus calcoaceticus

Bacterium anitratum

Acinetobacter anitratus

Achromobacter anitratum

Herellea vaginicola

Mima polymorpha

Acinetobacter lwoffii

Acinetobacter baumannii

and at least the *calcoaceticus* gives us credit for our unique ability to use calcium acetate as a sole carbon source. But when did a bacterium's opinion count with a human who wanted to impress or memorialize another? This renaming could go on forever, probably until the human race itself ends, or until microbes exterminate all taxonomists!

"Yes, my good chairbug, I realize that this all sounds very negative. I am merely venting a frustration that has been disrupting our metabolic equilibrium for many years (the only way we experience a psychosomatic dysfunction). I realize for us there is no easy way out. So, in retaliation for having our family name changed at least eight times, we suggest that the name *Homo sapiens* be changed to *Homo insensitivus*, as being more in keeping with the insensitivity many humans have for other creatures. I rest my case."

Acinetobacter, not being motile, descended from the podium twig and down the rock in a bouncing, bumpy motion, catching on small grains of sand to break its fall. All around, the metabolic activity of the assembled microbes rose abruptly in applause and discussion. *Escherichia*, in a flowing motion with its flagella flickering in the slanting rays of the sun, hastily again assumed the top spot on the podium twig and addressed the assembly.

"Fellow prokaryotes," it called out, waving its flagella wildly to regain the attention of the gathering. "Fellow prokaryotes! We now are ready to commence with the next full session of this congress—a review not only of the unique niches that we have occupied, but how some of these time-honored habitations have been altered, often to the detriment of human health."

As it spoke, Genie Transposon hopped up to the kiosk to set up a poster listing the proceedings for Session 1 in good view of the participants.

Session 1:

Habitats and Niches

Coleaders:
Legionella pneumophila and
Helicobacter pylori

Thermotoga maritima

Each of these colleagues will relate interesting stories of how they have emerged to strike temporary fear into the human race. Legionella's story is a prime example of how a peace-loving microbe, living contentedly in the waterways of the world, was suddenly transported to a new environment, with devastating effects on humans. Helicobacter will relate the exciting and unique story of how it was able for several decades to remain undiscovered in the hostile acid environment of the human stomach, also causing considerable human misery in the form of gastritis, gastric ulcers, and even cancer.

Advances in life's ability to encapsulate itself and move into new habitats can also be identified. The path begins with the first cells living in water. Then multicellular life emerged onto land, protected by coverings such as shells, skins, and bark, about 600 million years ago. The use of clothes and other artifacts enabled people to move to all climates, to establish cities on all continents within the last few thousand years and the move into the western "frontier" starting about 600 years ago.

—L. Margulis and D. Sagan, *Microcosmos*, 1986

Escherichia coli descended from the podium. *Legionella pneumophila* and *Helicobacter pylori* took their places, appearing together to introduce themselves to those assembled who had not encountered them before. These two colleagues were selected as coleaders for this session because of the notoriety each had gained from its unique habitat and relationship with human hosts, not always to the benefit of the latter. *Legionella pneumophila* was ready to open the session, while *Helicobacter pylori* settled back in its familiar ammonia cloud to wait its turn. After adjusting the top of the podium to accommodate its longer, slender length, as compared to the shorter, stubby form of *Escherichia coli*, *Legionella* was ready to address the assembly.

Legionella pneumophila's Story

"My good colleagues," *Legionella pneumophila* began, "we welcome you to this session. I plan first to review the different natural habitats that we prokaryotes have occupied for centuries. Each of you has a unique story to tell, but unfortunately limits of time preclude a full disclosure from each of you. However, time following the formal presentations of myself and my honored cochair, *Helicobacter pylori*, will remain open for select tales and sketches from many of you.

"The natural environments in which each of us can best thrive are usually orderly and peaceful until humans invade or expand our natural borders. It is amazing how humans can disrupt our time-honored habitats and have no sense of personal responsibility when such intrusions work to their disadvantage. It is my intent to exonerate us from any blame in these circumstances.

"Have humans not experienced similar catastrophes when they invaded or occupied pieces of land where they could best thrive? History records war after war where native insurgents or outside intruders would covet the same settlement, often resulting in many more human-caused fatalities than from any diseases with which we were associated. Elaborate means of protection—rock walls, castles with moats, swords, guns, and the assembly of armies—were among a variety of methods invented to protect one human enclave from the avarice of others. We cannot be held to blame for the emergence of infectious diseases that result from the alteration of our natural niches purely for human gain.

"Let me tell you the story of my family, to illustrate how changes in human practices led to outbreaks of an emerging new disease, 'legionellosis.' You may notice that my designated family name is linked to an initial outbreak of pneumonia among American Legion conventioneers, fondly referred to as 'Legionnaires.' Christening us with the name *Legionella* not only commemorates this event, but subtly carries a finger of blame as well. But I will let you decide, after I relate the following story, whether toying with our family name is justified.

"Members of my lineage have for centuries lived in the bottoms of natural lakes and quietly flowing streams. We have preferred inhabiting these mud bottoms primarily because we thrive best where the temperature is maintained between 28 and 30°C and where nutrients are in ready supply. The lake and stream bottoms are rich in iron, leached from the soil, and in cystine and other essential amino acids, extracted from decaying vegetation, that build our cells and maintain our metabolic functions. We must plead guilty to causing isolated cases of undiagnosed human pneumonia in the past, primarily among fishermen, swimmers, and other water sports enthusiasts, who may have swallowed or inhaled contaminated lake water. Then, in the summer of 1976, our notoriety was sealed. An explosive outbreak of pneumonia occurred among some old codgers at the annual convention of the American Legion, held at the Bellevue Stratford Hotel, an edifice romanticized as the 'Grand Old Dame of Philadelphia.' An unexpected outbreak of 182 cases of pneumonia, of sudden onset and often fulminant, turned terrifying when 29 of the Legionnaires died.

"From our perspective, and from the perspective of many other prokaryotes who were sideline spectators, the evolution of this 'new' disease can be attributed entirely to the folly and self-indulgence of humans. Humans do not like to sweat. Sweating is a purely natural response to keeping the body cool, but it is uncomfortable and holds the added stigma of emitting unpleasant odors, for which a variety of antiperspirants have been invented. How do humans protect themselves from sweating? Air conditioning! How is the air 'conditioned'? Heat is transferred from a body of air to huge water vats known as cooling towers. And where does the water in the cooling towers come from? You guessed it: from nearby lakes and streams, where the water is not adequately filtered to prevent the entrance of our clan and other bacteria. Now you have the picture. Humans have created a second, almost utopian environment—warm and stagnant—in which our progeny can thrive. Although we're quite content with our niche in natural waters, a new world in the controlled temperature of air conditioning cooling towers has been introduced to us, where we are less threatened by the adverse changes that can occur in our natural environment.

"Before 1976, we remained hidden in our new world, largely unrecognized by humans, or, for the few who might have known, of little interest. That is, until the next part of the story evolved.

"In central air conditioning systems, the cooled air is directed into a vast duct system that is distributed throughout the building. Once again, this conditioned air is not filtered, not to the point of preventing our access. We ride along with the dewdrops of humidified air that are delivered throughout this duct system. And we find a whole variety of new habitats in various nooks and crannies throughout the air-conditioned building.

"As the next step in this saga, I shall relate yet another sphere of human progress, or folly. Humans also do not like to stay in one locale (*Homo mobilis* perhaps should be their new name) and thus have invented several means for transportation. In this modern era, transportation is so convenient that even the elderly and debilitated can find conveyances: even the creaky old soldiers who in this story elected to join the American Legion. And further, they take advantage of buses, trains, and airplanes to attend conventions together to promote good will and camaraderie. You can imagine them ambulating about with their canes, huffing and puffing up elevators and escalators to ballrooms and meeting rooms to join in the festivities. Where do you suppose these old soldiers slept? You got it! Right under the air conditioning ducts in their hotel rooms! And now my story is a bit less mysterious and more predictable.

There we were, before anyone knew who we were—along with other microbes that have also found air conditioning and water delivery systems to be ideal places for habitation—lying in wait to explore the next phase of our odyssey. The cool mists, on which we rode like mythical Arabians on flying carpets, were directly inhaled into the respiratory passages of those unsuspecting, sleeping veterans.

"Now imagine! We found ourselves in a world even more glorious than the cooling towers. The moist confines of the bronchial tree were simply the best environment we ever had! From the bronchi we were transported upstream into the bronchioles and finally into the remote alveoli or air sacs in the lungs. This was almost like being back in the muck of our river bottoms. Warmth, moisture, and nutrients were in perfect combination. To their disadvantage, the often compromised health of these old soldiers left them with virtually no ability to fight back. We easily dodged the few scavenger phagocytes that circulated in our midst, and the feeble effects of occasional antibody molecules posed little problem for us. We did only what comes naturally—we grew and grew like dandelions!

"Our unchecked growth, and the minimal reaction these old soldiers could muster, resulted in a condition that humans call pneumonia—and, in dubious honor of the Bellevue Stratford Hotel outbreak where we were first hunted down and identified, 'Legionnaires' disease,' from which our name, *Legionella,* was derived. In our case, the human doctors use the term 'lobar pneumonia' or, even more dramatically, 'double pneumonia,' when our reproductive prowess really gets into high gear and we bridge across the mediastinal Alps into the opposite lung. In fact, our extended name, *pneumophila* ('loving the lungs'), was coined to acknowledge this special predisposition.

"In some instances our activities became too zealous, so that these old soldiers not only faded away but died, an outcome to neither their nor our best advantage. For you see, any human who succumbs represents suicide for us as well, an eventuality unforeseen by us in such a bull market environment. Investigators later disclosed that the conventioneers most susceptible to our ravages were the very elderly or those who smoked, drank, or partied too much (indulgences that further stressed their already marginal immune systems).

"The prolonged length of time it took investigators to discover that we were at the bottom of all this ruckus is of greatest surprise to us, particularly in this advanced age when humans flaunt their intellectual sophistication and scientific acumen. 'Swine flu,' pontificated the chairperson of a government task force. 'Poison

Gas,' or 'Biological Attack by Russian Spy Agents,' guessed the newspaper headlines. Even some of the crack microbiologists of the time could offer nothing better than promoting a 'virus,' an 'unknown microorganism,' or a 'lethal toxin' as the cause of this plague. Heavy metals or allergens in the water were other theories. This desperation to find a cause was little better than that of the 14th-century Council of Venice that was truly shooting in the dark to ward off the Black Death, as so eloquently related by *Yersinia pestis* in the last session.

"Our initial mass attack occurred during the summer of 1976. Not until late in the next spring did researchers, somewhat by accident, finally extract us from our comfortable new homes. We first were enticed to duplicate prodigiously in laboratory culture, in chicken eggs of all things. You see, we are not capable of growing in most of the bacterial culture media being used at that time. Most culture media contain toxins that we simply cannot tolerate. Humans call us fastidious, but can we help it that we need iron, cystine, serine, methionine, arginine, valine, leucine, isoleucine, and threonine, among other nutrients, just to survive? When none of these essential nutrients are in the culture media in common usage, we are no better off than old-time sailors with scurvy.

"Finally and probably inevitably, after many blind attempts, the combination of charcoal to adsorb the noxious substances from the culture medium, along with extracts of yeast cells which contain virtually all of the nutrients we need to reproduce, led to the scientists' success. Lo and behold! There we were, peering out with iridescent beauty from the black fog of this newly discovered yeast-charcoal culture medium.

"In closing, let me tell you of an adaptive contract we have with nature which has also caused humans much frustration. Our eukaryotic friends the amebas (a race far advanced over us because they have a nucleus) share many environments with us, including our natural water habitats. In fact, certain amebas make it a habit to graze on bacteria. This is obviously not always to our advantage, but we have made a symbiotic accommodation with them. Once ingested inside their cytoplasm, we surround our-

The Other End of the Microscope

selves with a membrane-lined vacuole in which we not only survive but are able to reproduce. The amebas provide us with protection, and in turn we manufacture a variety of chemical products that they use for their internal metabolism. What a fair arrangement!

"But do you see the implications of this strange quirk in nature? Once humans learned who we were, lurking there in their air conditioning ducts, every measure was put in force to eradicate us from the scene. They attacked us with hot steam and vapors and threatened us with all sorts of chemicals. 'Decontamination' is the word they use. After a series of such treatments, they cultured material from the inside of the ducts. 'No growth': we weren't there. Sterile! Kaput! Cheers rang out, and glasses were raised high in toasts of victory!

"Following a well-planned scientific protocol—or perhaps feeling a bit cautious after their hubris—the scientists cultured the air conditioning ducts again a few weeks later. Agony! There we were!—iridescent colonies proudly growing on their charcoal agar plates. It took many additional months for those humans to realize that we had holed up within our friendly amebas, which protected us from their methods of mass extermination. Once the storm was over, we were regurgitated hale and hearty from the surviving amebas and resumed our happy life in the mild waters of the air conditioning towers. Humans will spend the rest of their days being fooled by us simple creatures, and will go to their graves trying to find all the nooks and crannies in which we microbes can hide, ready to sally forth at the next act of human fallibility.

"In summary, my fellow bacteria, we would have been content to remain in the lakes and rivers of the world as we have done since time immemorial. We should not be held responsible for causing humans suffering that has resulted purely from their craving for body comforts. We are the victimized, displaced from our natural habitat and forced to live in strange and distant locales. Not that our deportations did not open up new worlds for us, for which I suppose we should be grateful. But in reality, we are only simple folks, following what comes naturally as encoded by the genes that reside in our chromosomes. We cannot be held responsible for the diseases that result when humans interject themselves into our homes or create new environments in which we thrive. A plea to all humans: leave us in the natural waters where we are most comfortable, where we do not need to counterattack—we have no innate desire to cause anyone misery!

"Fellow prokaryotes, I thank you for your attention. I will now pass the baton to my friend and colleague *Helicobacter pylori*, who will tell you an equally amazing story of survival and how it has only recently been recognized as a surprising link in the causes of a time-honored human condition."

Spontaneous Effusions

At the conclusion of this discourse, a noticeable acceleration of the growth curve was detected among the attendees of this session. *Legionella pneumophila* was thus assured that its presentation had gone quite well and that many new perspectives had been gained on the human tendency to blame the prokaryotes for the results of human error.

Just before *Legionella* jumped back onto the pond, and as *Helicobacter pylori* was ascending the rostrum, *Mycobacterium marinum*, reflecting sunlight off its yellow, shiny, waxy coat, signaled that it would like to interject a comment. *Legionella pneumophila* waved in assent.

"I wish to commend you, *Legionella*," announced *Mycobacterium marinum*, "for a most interesting and informative discourse. I and my mycobacterial ancestors also have shared in your habitat of the waterways of the world, although you and we may not have crossed paths that often. We much prefer the seaside and the salt-rich estuaries. We are not motile and therefore would have difficulty in keeping up with you. Besides, our waxy mantle makes us less dense than water, so that we hover near the surface rather than at the bottoms of streams. We share with you, however, in the human category of being 'fastidious,' although such a condition has never been stressed in our chromosomal codes to the point of necessitating a mutation. Everything we needed to carry on our business was provided in our natural salty homes. But I have just a brief story about an unusual niche in which we once found ourselves.

"In the Rocky Mountain West, near a region they call the Grand Mesa, is located a hot mineral spa, so situated to be of delight to both us and our human friends. Talk about nutrients—better than any seawater we ever found! All those hard-to-come-by minerals were there in abundance in this spa. I am sure that our usually slow doubling times were cut at least in half, just living in that piece of paradise. We grew so fast that we experienced our own style of urban sprawl.

"One day a group of humans found our niche and decided to build a swimming pool. They dug out our burrows and lined the whole pond with a thick layer of cement. In order to ensure that swimmers did not slip and hurt themselves while frolicking in these healing waters, they made the cement surfaces rather rough. Success—except the swimmers acquired a variety of scrapes and scratches on their knees and elbows. Not knowing any better, some of us decided to take up housekeeping in these skin abrasions. We made out very well in our newly found niche, where the constant skin temperature of 32°C and the physiologic balance of salts suited us to a tee.

"Not long afterward, however, we found ourselves in a battle for survival with a whole host of human defensive cells—hordes of mononuclear and multinucleated scavengers invading our new homes. In very short notice, all the waste carnage from the conflict resulted in the formation of distinct lumps and bumps beneath the infected humans' skin. Being always innovative with new terms, doctors designated this new condition 'swimming pool granuloma.' Picking up on one of your comments, *Legionella*, we also were surprised how long it took humans to figure out that we were the source of this leprous condition. It took many tries before they finally found us. Literally hundreds of culture medium formulas were tried, without success. We grow optimally at 30 to 32°C (the temperature of skin) and little if at all at 35 to 37°C, the conventional temperature of their incubation chambers. Once they discovered our optimum temperature, the formula for a new culture medium was finally derived. In retrospect, it is interesting to speculate that the contents of this new medium should have been more predictable, as the composition of the new formula approximated the mineral content of the spa water from which we had just emigrated. After that it was all downhill for us. They brought out the powerful antimicrobial guns, and it was not long before we were evicted."

From its perch on the podium, *Helicobacter pylori* looked out over the assembly, noting the approving efflux of gases being emitted. "Thank you, *Mycobacterium marinum*, for this intriguing story. It seems to be well received." Then, with one last undulating, corkscrew motion, *Helicobacter pylori* positioned itself for its own discourse.

Helicobacter pylori's Story

> The extensive genetic diversity between isolates within a single *Helicobacter pylori* species, taken together with the substantial genetically determined differences that have been observed among separate *Helicobacter* species, suggests that many more insights will be needed before the current, medically inspired curiosity surrounding these organisms can be satisfied.
>
> —Jeffrey L. Fox, *ASM News*, December, 1999

"Fellow prokaryotes. In keeping with the puzzle over why humans, with all of their scientific knowledge and ability to interchange information, took so long to link *Legionella pneumophila* with the old Legionnaires' pneumonia syndrome, and, as we just heard from *Mycobacterium marinum*, the confusion surrounding the 'swimming pool granuloma' mystery, the story that I am about to relate will be equally amazing for the curious among you.

"Decades passed before humans realized that members of my clan played an integral part in the evolution of peptic ulcer disease. Using their marvelous intuition and empirical deductions, until only recently they targeted virtually everything imaginable as causes of this worldwide affliction. Nerves, undue stress, or ingesting too much chocolate, alcohol, or caffeine were among several particulars blamed for the affliction. 'Slowing down,' 'keeping one's temperament under control,' and 'taking time to smell the roses,' so to speak, have been prescribed as prerequisite to prevention and healing. Excess gastric acid in an agitated stomach was believed to literally burn holes in the gastric and intestinal linings. The observation that symptoms were less when the stomach was full of food and most irritating when hunger was at its peak substantiated this point of view.

"But they couldn't explain why peptic ulcers developed in relatively calm and cool folks, in whom they could demonstrate no irregular patterns of gastric acid secretion. Well, it never takes *Homo sapiens* long to posit a number of theories, some of which are a bit elaborate and understood only by those high on the academic ladder. Mucosal resistance seemed to be the best bet. You see, the cells lining the stomach must remain healthy and intact to resist the eroding action of gastric acid. Some thought that alteration in blood supply was involved in ulcerogenesis (a fancy word that could only have been invented by a human who was trying to impress a professor or colleague). Another person even attempted to demonstrate that peptic ulcer subjects did not replenish their intestinal epithelial cells at the normal rate. 'Normal rate,' measured by ingenious methods never made entirely clear, was found to be an average of 4 days. Sophisticated terms such as 'leak-back rate' and 'back-diffusion' of hydrogen ions were advanced to explain differences in susceptibility in response to the action of gastric acids. The ingestion of alcohol, as in schnapps among other libations, or of salicylates such as aspirin, was for several decades considered a major 'barrier breaker,' specifically to be avoided by people not wanting their stomachs 'eaten out.'

"Many years ago, one group of investigators came quite close to discovering us as the true cause of peptic ulcers. You see, spiral bacteria had been observed in the biopsies of human stomach linings as far back as the turn of the 20th century, but no one paid much attention. These observers never guessed at us helicobacters as one of the chief culprits causing gastritis. We were merely considered to be innocent bystanders, simply curiosities; that is, until the early 1980s, when two guys in Australia, J. R. Warren and Barry Marshall were their names, finally found the combination of ingredients, the ideal incubation atmosphere, and the optimal temperature to recover us in cultures. From that time on we were doomed, subject to all modes of human investigation and attack. We must give credit to our human friends, however, for not attaching their own names to our race. It is surprising that we did not end up being known as *Marshallobacter warrenii* or some other such appellation.

"We were temporarily agitated when they first named us *Campylobacter*, linking us with a group of bacteria that really are not at our level of sophistication, living on the 'other side of the tracks,' as humans would say. This really is nothing against the campylobacters, but, you see, they live downstream in the intestine, where living conditions really do get a bit mucky. I will reduce my effusion momentarily to let you here assembled in on a little-known secret. Dr. Warren, from his exacting studies, knew all along that our unique attributes were far too sophisticated to be included with the campylobacters. However, during the year of our discovery, he hoped to attend an international *Campylobacter* conference in Stockholm, Sweden. So he published the characteristics of a 'new' *Campylobacter* species, uniquely linked to peptic ulcer disease, and an invitation indeed arrived. You can imagine the widespread excitement that this announcement elicited among the scientific community. Not since that day when Robert Koch announced 'old tuberculin' as the cure for tuberculosis had such excitement been generated. Medical practitioners literally flocked to hear his '*Campylobacter*' story. Later, when it was evident that we helicobacters in reality did not belong to that group, taxonomists had the courtesy to name us after one of our unique attributes—the wiggly, corkscrew, helical-like nature of our physiognomy. The designation *pylori*, positioning us in one of our major homelands, is accurate, though a bit less alluring.

"What makes us so unusual and worthy of a story? First of all, we have special complexes on the outside of our cell membranes (humans call them receptors) that allow us to attach to the surfaces of the cells lining the stomach. Most of my bacterial colleagues would look at this as a death knell, as exposure of even a few seconds to the extremely low acid pH of the stomach environment can quickly coagulate our innards. Most bacteria that are ingested with water or food must make a rapid two-step through the stomach, or hide within some well-buffered, protein-rich morsel during transit, to thwart the action of gastric acid. But our clan has been endowed with a special means of survival in this most hostile of environments, of which humans have only recently become aware. We share with only a few of our comrades the ability to break down urea rapidly. Those among you who have the rudiments of chemical knowledge know that when urea is broken down, ammonia is formed. Further, as you will deduce, ammonia is alkaline, providing an effective neutralizer to the acid hydrogen ions that are in such high concentration in the stomach.

"Now, let me summarize the situation. As I have explained, we can attach to the surface of a stomach epithelial cell, wriggle with our flagella, and move corkscrew-like under the surface layer of mucus. This provides us with the momentary protection we need until sufficient urea is hydrolyzed to surround ourselves with an 'alkaline cloud.' Once this cloud is established, even the strong gastric acid can be buffered, allowing us to survive within this cocoon. By establishing points of connection on the epithelial cells, we inactivate the microvilli, and the increasing concentration of ammonia results in direct damage to the mucosal cells.

"Once we are able to establish a foothold, other mechanisms which the humans call 'virulence factors' are brought into play. Our ability to produce proteases and cytotoxins further damages

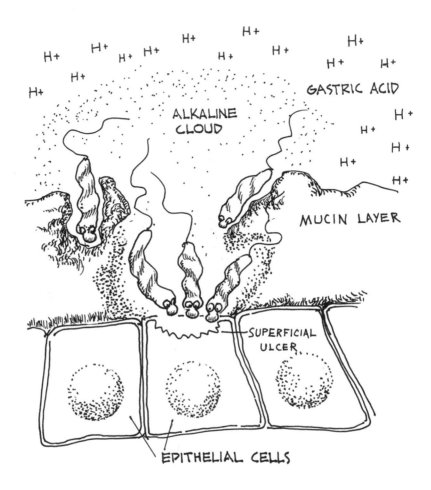

H+ H+ H+ H+ H+ H+ H+ H+ H+ H+ H+ H+

GASTRIC ACID

ALKALINE
CLOUD

H+ H+ H+ H+

MUCIN LAYER

SUPERFICIAL
ULCER

EPITHELIAL CELLS

the mucosal cells. In addition, we produce substances that cause an influx of phagocytic cells, resulting in a dense inflammatory response, or gastritis. Taken together, these changes make the gastric epithelial surface more susceptible to the digestive action of gastric acid, which in combination ultimately results in the formation of an ulcer. In the ulcer bed, the damaged and disintegrating mucosal cells release carbon and nitrogen compounds in our immediate surroundings, a milieu ideal for our rapid replication, providing a beachhead from which to continue our assault. One can readily understand why humans are so bent on reversing this snowball effect.

"In the early days of antiulcer therapy, the extensive use of antacids was among our more pleasant surprises. The antacid had the effect of neutralizing the gastric acid even more, providing us with an environment even more conducive to our success. More recently, however, bismuth was added to the gastric ulcer regimen, administered along with combinations of new antibiotics. This stuff is as lethal to us as strychnine is to humans. We are now under siege, which in many instances has resulted in great destruction among us, at times eliminating entire enclaves. Humans have called this onslaught against our clan a 'cure'; this, of course, depends on one's point of view. Humans should not be too complacent. For you see, it is only a matter of time before we helicobacters readjust our internal mechanisms of defense. We will find ways of rendering ineffective the actions of bismuth or the prolonged use of any antibiotic. Even if only a few of us survive, we can always establish new enclaves, retooling for a renewed and even more vigorous attack on the gastric mucosa.

"So, this is my story. We now have a few moments remaining for any of you here assembled to make comments, or share briefly your individual niches of existence that may be of particular interest."

Spontaneous Effusions

Legionella pneumophila again ascended to the podium and, along with *Helicobacter pylori,* looked out over those assembled to receive any comments. *Pseudomonas fluorescens* flashed its yellow light and prepared to speak.

"Colleagues of this assembly," began *Pseudomonas fluorescens,* as its highly active polar flagella allowed its quick transit to the podium. "These discussions are highly enlightening. I have one comment. Humans should realize that various prokaryotes occupy exotic environments far beyond any realms in which humans could exist. Although men and women can adapt to living in the tropics or on the Arctic Circle, along the seashore or high in the mountains, we not only are omnipresent there, but extend to the borders of life's extremities. For example, our close cousin *Pseudomonas aeruginosa* has adapted to living at increased temperatures, up to 42°C or even 45°C for some strains. This property is not all that impressive compared to what *Thermotoga maritima* has related, but it is useful enough. Many of us prokaryotes, called thermophiles, content ourselves with occupying any niche where heated water accumulates for any length of time. Look around the hospital environment: hot tubs, rehabilitation pools, nebulizers, and the like. You are apt to find us lurking there. It should be no surprise when a skin or soft tissue infection develops in a person with an open wound, particularly one who has just used a thermal therapy tub or who has recently frequented a contaminated hot water spa. In contrast to the swimming pool granulomas described by *Mycobacterium marinum*, we are much more aggressive, producing large areas in which the skin simply sloughs off.

"In reference to thermophiles, others of our heat-resistant colleagues, such as *Sulfolobus acidocaldarius*, dwell in the hot springs in Yellowstone Park, where the temperature reaches 80 to 85°C. Others can even grow in deep ocean volcano smokers where temperatures exceed 100°C, the boiling temperature for water. Discounting complete destruction of planet Earth as we know it, we prokaryotes will survive any mass event, be it thermal or nuclear, that potentially can annihilate the human race. Given millions of years of future evolution, out of these ancestral remnants we prokaryotes may even develop an entire new race of multicelled eukaryotes."

"You make a good point, friend *fluorescens*," Helicobacter returned. "Are there any other comments?"

"Yes. *Yersinia enterocolitica* here. On the opposite end of the thermal spectrum, we have the ability to survive and replicate at very low temperatures. Thus, within any nutrient substance that humans may elect to store at refrigerator temperatures, we may be hanging around. Our ability to survive and replicate in refrigerator-stored transfusion blood is one prime example where we cause a bit of a fuss. You can imagine the celebration we have when released into the warm climate of a transfusion recipient's bloodstream after being in the cold for so long. Upon this release we absolutely go crazy, though we are soon discovered as our elation causes chills and spiking temperatures in the blood recipient. To be sure, humans can understand this response to escape. Ever see how they frolic about after a prolonged cold snap? Same thing!"

"Good point, *Yersinia*," chimed in *Bacillus cereus* from a point at the rear perimeter. "We also like to hide out in stored foods, particularly in oily foods. We can't go quite as low in temperature as you have described, *Yersinia*, but we can reproduce very nicely at room temperature. When humans leave us in such an environment for any length of time, we get into their craw, literally speaking. Leave us in a bowl of rice long enough and we produce potent toxins. We ourselves, thanks to our gentle nature, have no direct effect on the gastrointestinal lining when we're ingested, but our accompanying toxin can create real havoc. Our unsuspecting human begins to retch and vomit everything eaten. Nasty business—everyone who tried the salad will be affected, pretty much erasing any pleasant memories of the picnic."

"Yes indeed, *Bacillus cereus*," answered *Legionella* from the podium. "But I would like to reiterate one point that I made during the telling of my story. Humans will blame you for what they call 'food poisoning.' Am I correct? You are held to blame? Of course you are. But in reality, the afflicted humans first of all used unclean containers and/or implements or did not properly wash their hands, letting you into the food in the first place. Then they failed to keep their culinary creation at temperatures below which most 'food poisoning bacteria' can grow. How can you be held liable for their oversights? I do not believe you have any cause for self-flagellation for merely carrying out your creative nature."

Helicobacter pylori, sensing that its exponential growth was about to convert to the stationary phase indicating exhaustion, descended from the podium, making way for *Enterobacter aerogenes* and *Micrococcus luteus* to open up the next session. During this brief interval, Genie Transposon placed the next session's agenda on the kiosk.

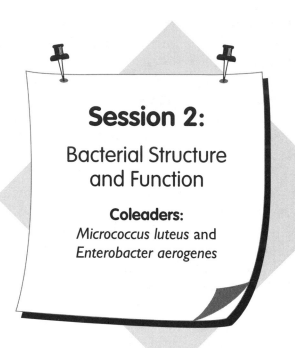

Session 2:

Bacterial Structure and Function

Coleaders:
Micrococcus luteus and
Enterobacter aerogenes

Thermotoga maritima

These honored bacteria, Micrococcus luteus *and* Enterobacter aerogenes, *were selected as coleaders because they represent two distinct cell wall structures, one called "gram positive," the other "gram negative," for reasons to be elucidated. Each will demonstrate that bacterial metabolism is not simple; in fact, it is equally as complex as the biochemical reactions taking place in any single human cell. The one shortcoming of the bacteria, to date, has been their inability to assemble and organize into a functional organ, although their aggregation into distinct colonies and the formation of a glycocalyx or biofilm, to be described in a later session, provide a means for collective metabolic activities unique to each species.*

As the subject of bacterial anatomy has the reputation of being a bit dull and boring, those here assembled were in for a bit of a surprise. Micrococcus luteus, *in particular, has much to say about the wonders of the bacterial structure, which is only slightly less complex than that of individual cells in higher plants and animals. An uninitiated observer, gazing at a stained preparation of bacteria through a microscope, might disregard as insignificant the fragments of material in the field of view that superficially resemble bits of dust. However, a closer look through a scanning or regular electron microscope will reveal structures that mimic all other cells in existence. In this regard, can you even fathom how a somewhat obscure and little-read classical 17th-century German philosopher, Gottfried Leibniz, could have had us intuitively in mind even before anyone knew we existed? He wrote,*

> The universe is composed of a hierarchy of monads each of which is a microcosm reflecting the world with differing degrees of clarity from its particular point of view in a system of harmony pre-established by God.

Pseudomonas *will have more to say in another session about the incorporation of the word* monad *into its name. Nonetheless, the prokaryotes should adopt Leibniz as their patron saint. How could their existence and reality be better stated? Tiny and obscure as they are, each species participates in its own unique niche in the universe, in a microcosm equally as significant as that of other cells, while reflecting the world from its own singular point of view. This is what this treatise is all about.*

Enterobacter aerogenes, being freely motile via its peritrichous flagella, easily ascended the podium; its copresenter, *Micrococcus luteus*, being a sphere, more slowly rolled up the rocky crag. Amid the varied colonies, the smooth, dome-shaped, yellow mass out from which *Micrococcus luteus* ascended was more than impressive and drew the close attention of those in the audience. The session on bacterial structure and function was about to commence.

Presentation of *Micrococcus luteus:* The Wonders of Bacterial Structure

"Welcome, colleagues. It is good that you could come, as we have important matters to discuss in our relationship with the rest of the world, particularly *Homo sapiens.* Please bear with me momentarily as I review a few technical terms, to be sure we all start at the same point of reference.

"I, *Micrococcus luteus,* represent a large group of prokaryotes that humans call 'gram-positive cocci'—I will explain presently how they came up with this term. Sorting out the prokaryotes into structural and functional categories has been a relatively recent task for them, as we have been recognized for only a little over 100 years. Of course, humans have since inception categorized their own kind—where they live (Europeans, Africans, Chinese, etc.), how they look (Caucasians, Asians, blacks, and so on), the religion to which they belong (Christian, Muslim, Buddhist; or, even more refined, Catholic, Episcopalian, Presbyterian, etc.), or the ideology they profess (Democrat, Republican, Conservative, Libertarian, 'Green,' or whatever). Thus, we perhaps should not be disturbed that humans also want to arrange us prokaryotes into some categorical hierarchy based on how we look and how we function. Some among you may be distressed that we are often tagged with nondescript alphanumeric designations such as O157:H7. But even humans may similarly carry around impersonal identification tags—Social Security numbers, dog tags, personal identification numbers ('PINs'), or a plethora of telephone numbers and e-mail addresses!

"To begin with appearances: members of my clan are known as 'cocci,' simply because we are round or spherical in outline. We are further classified based on how we arrange ourselves. My brethren who collect in clusters are called staphylococci, from the Greek stem *staphylo,* referring to 'a bunch of grapes.' Those who prefer to line up in chains are called streptococci (*strepto,* meaning 'twisted' in Greek); our friends that prefer not to split upon division are colloquially called diplococci. But such descriptors should not surprise you. Humans gather like bunches of grapes in stadiums, line up in rows or chains when waiting for something, or pair off for purposes of companionship and reproduction. My specific clan, *Micrococcus*, prefer to arrange in groups of four (which humans refer to as 'tetrads'), which also is not inconsistent with those humans who gather to play four-handed bridge or team tennis or dance a quadrille. It would seem that prokaryotes and humans are much alike in many traits!

"In fact, we cocci have a fundamental, scientific reason for arranging the way we do—which perhaps is not as clear cut for humans. The diplococci and the streptococci have a set of enzymes that allow division only in one plane; therefore, when reproducing, they string out in pairs or long chains. In contrast, my fellow micrococci and staphylococci possess a second set of enzymes that allow us to divide in a second plane; thus, one can visualize a sphere quartered, and each part remaining in tetrads or aggregating into grapelike clusters. Perhaps, now that the human genome has been elucidated, some bright researcher may find that the human predisposition toward gregariousness and an innate desire to participate in group activities may also be coded in their DNA (some work is indeed currently under way to determine if there may be a genetic predisposition separating introverts from extroverts). But I doubt this breed of humans will ever be called *Staphylohomo*!"

Laughter rippled through the audience, in an amused release of metabolic gases. *Micrococcus luteus* bowed, and continued as the cloud of mirth drifted away across the pond.

"Now, about this gram-positive and gram-negative business. Much later in the evolution of the human exposure to the prokaryotes, along came a fellow by the name of Hans Christian Gram (not Anderson, or Grimm, both of fairy tale fame), who discovered a better way for humans to look at us. He dipped us in some crystal violet dye, which we happily soaked up; he then sprinkled us with aqueous iodine to cement the pigment inside us. Finally, Dr. Gram gave us a shower with acetone-alcohol in an attempt to rinse out the dye. Half drunk after this treatment, we

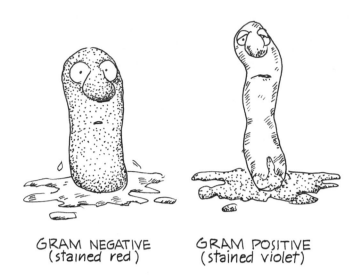

GRAM NEGATIVE
(stained red)

GRAM POSITIVE
(stained violet)

were finally dipped into a red dye, safranin, that humans call a 'counterstain,' and this is the process as it is used even today. The whole procedure is almost like being run through a car wash. This rough treatment actually has little effect on us beyond stopping our activity—or, more precisely, painting us into a corner.

"Those of us who retain Dr. Gram's violet dye, and appear blue-purple in color, after the alcohol douche are termed by humans 'gram positive.' In contrast, our colleagues who don't absorb the crystal violet—who freely let it go in their alcohol shower and eventually instead absorb the red safranin counterstain, resulting in a pink color—these are called 'gram negative,' a descriptive rather than a reflection of any personal traits. So my *Micrococcus* cousins and I, being spherical and dyeing blue in the Gram stain, are termed 'gram-positive cocci.'

"The reason for this difference in Gram stain reaction is because of our outer structure. All of us gram-positive prokaryotes are encased in thick cell walls. Our ancestors devised this strategy of a rigid wall to protect against the powerful forces of osmosis. In any situation where more particles are on one side of a semipermeable membrane that allows passage of only water molecules, the flow of water will be in the direction of the chamber with the denser number of particles: in this case, into our cytoplasm. Without our rigid cell wall, water would flow into our particle-rich interior until we blew up like a balloon and burst. (Now, you may have heard of certain of our cousins, the 'protoplasts,' who exist without a cell wall. These brave fellows, however, can survive only in environments where the concentration of particles on the outside of their cells is high, such as occurs in urine or pus-filled exudates. In such environments, the number of particles outside the cell membrane is practically equal to the number inside, thus balancing the osmotic diffusion of water.)

"Our cell walls are composed of peptidoglycan, a cumbersome word that refers to the proteins (amino acids) and carbohydrates (sugars) that serve as our base chemicals. The carbohydrate part of our cell wall can be likened to the 2-by-4 timbers of a frame house. These structures, called disaccharides, are composed of two sugars called *N*-acetylglucosamine and *N*-acetylmuramic acid, held together in tandem through an oxygen (glycosyl) bond. This long, linear sugar chain can be visualized as the long studs in our mythical frame house.

"However, in contrast to the frame of a human house, in which supporting cross braces can be placed anywhere at will, the cross links in our cell wall are limited to special connectors within the muramic acid moiety. Muramic acid is unique to prokaryotes. It is linked to chains of amino acids, within which alanine and lysine serve as the hinges to which are latched the cross supporting braces. (These cross links for us gram-positive bacteria include a string of five glycine molecules; for our gram-negative friends, they are made up of diaminopimelic acid residues. These details are not important for our discussion here, but will come into play in a later session, when we plan to discuss the action of antibiotics against the construction of cell walls.)

"With the gram-positive bacteria, this grid formation proceeds in three dimensions, in the end producing a rigid, four-cornered model known as a 'tessera' (*tessares* in Greek meaning 'four' or 'four cornered'). Ultimately our entire surface is covered with layer after layer of carbohydrate struts supported by protein cross braces, just like the walls of a house. In contrast, the peptidoglycan cell wall of our gram-negative cousins is very thin, consisting of only a single layer that offers much the same degree of protection that we enjoy. Again, more of this later.

"Now, not only does this thick cell wall provide for support against osmotic forces, it also serves as a barrier against physical agents and penetration by harmful substances, including some antibacterial chemicals. Although rigid, however, the cell wall is porous, pierced by tiny holes much like ventilation in a house, which allows certain substances, such as the crystal violet dye particles in Gram's stain, to percolate into the space (called the periplasmic space) between our cytoplasmic membrane and the cell wall. Inside the periplasmic space, the dye combines with the iodine 'mordant' to form crystals that are too large to get back out through the cell wall on their own. Because the gram-positive cell wall is so thick, the alcohol wash dehydrates it, closing its pores and blocking the alcohol from reaching the iodine-dye complex and washing it away.

"On the other hand, my colleague *Enterobacter aerogenes*, as representative of the gram-negative bacteria, will relate in a moment why structural differences in its much thinner cell wall allow it to release the crystal violet dye when 'under the influence' of the alcohol douche.

"So much for our outer structures. Inside, all of us, both gram-positive and gram-negative bacteria, have a cytoplasmic membrane that encloses our jellylike inner cytoplasm. In this endoplasm are trapped enzymes, various positively and negatively charged ions, food and energy-rich granules, and other important inclusions.

"First of all, we are rather proud of our sophistication in having a chromosome. It is a bit different from what is found in the nucleated eukaryotic cells, but it serves the same function. Our chromosome is not contained within a nuclear membrane (after all, we are prokaryotes—lacking nuclei, that is), but rather floats free in the cytoplasm, or may be attached at one or more sites to the inner cytoplasmic membrane. Our chromosome (genome), called a nucleoid, is double-stranded DNA, constructed in a closed circle and wound in a supercoiled structure much like a tightly twisted rubber band. If this chromosome were stretched out, it would exceed the length of an individual bacterial cell by 1,000 times or more! So we have a huge bank of codes out of which to operate, and to mutate.

"We also possess ribosomes, which you may envision as the miniature factories that produce the long chains of proteins we need to function. Ribosomes are made up of two subunits of ribonucleic acid (RNA), large and small. Each subunit is composed of many proteins assembled onto a folded backbone of ribosomal RNA, or rRNA. The ribosomal subunits combine to form a machine for constructing protein strands out of the 20 possible amino acid building blocks.

"Now, you can imagine each amino acid in a protein as like a single pearl in a necklace. The order of the pearls on the necklace determines the protein's shape and function. This all-important sequence of pearls is predetermined by a messenger RNA (mRNA) molecule, like a jeweler's design. The mRNA molecule is a copy of a gene coded in our DNA; almost all the genes in our DNA are patterns for making particular proteins.

"The mRNA strand is held in a cavity in the ribosome, where the three-letter 'words' in its design or message (each word signifying one of the 20 amino acids, or a full stop) are translated into the language of proteins by transfer RNAs (tRNAs). Each tRNA carries a particular amino acid, and it latches onto its appropriate word in the mRNA message. The ribosomes then function to clip off the new amino acid pearl from its tRNA and link it to the previous pearl in the protein necklace. This process repeats until all the pearls specified in the message have been added, and the completed protein is then released. Voilà! A new protein!

"Thus, all the proteins produced by any given bacterium are preprogrammed—enzymes, toxins, and other substances responsible for the observed or phenotypic properties by which each of our species can be characterized. Though the 70S ribosomes of bacteria are smaller and subtly different from the 80S ribosomes of eukaryotes, the mechanism of protein production I have just described is precisely the manner in which the individual cells of plants, animals, and other eukaryotes also carry out their internal affairs.

"We prokaryotes differ from eukaryotic cells, however, in that our cells commonly contain plasmids, which are miniature versions of our chromosomes. We also can come under attack by bacteriophages, viruses that infect bacteria by injecting their own DNA material into the cell. In some instances, viral DNA may become an actual part of our chromosome, as a 'prophage'; more commonly, it's employed in manufacturing new viruses.

"Transposons too can affect the DNA message. These mobile segments of DNA can hop from one part of the chromosome to another, altering the genetic code and extending the repertoire of protein-coding genes. Often this enables the production of toxins and other noxious agents that are not normally part of our nature. What at one moment is a peaceful symbiote may, under the influence of a deadly plasmid, be converted into a virulent pathogen, inflicting disease on any plant or animal susceptible to its virulence factors.

"Nevertheless, plasmids are among our best friends and often serve to preserve our integrity in the face of antibiotic attacks. These circular bits of extraneous DNA, assimilated from neighboring bugs, are far more than stray fragments that merely wander in and out of our existence. They may be a link at the very threshold of life itself.

"First of all, they are capable of self-replication once they have been taken into our cytoplasm. Plasmids are adaptive to the micro-ecosystem of a new host. They can acquire new genes, rearrange old ones, and retain genetic information best adapted to the conditions of their new hosts. Thus, they can sort themselves out to ensure their survival, often independent of the survival of the host cell itself. Upon death of the host cell, the released plasmid may find a new host, carrying along its newly developed abilities.

"You probably have already deduced the significance of this phenomenon. If there developed in one of us a set of plasmid-related genes that encoded resistance, by whatever mechanism, to the action of an antibiotic, these genes could be transferred intact via conjugation to our next of kin or even to distant cousins, making them resistant to that antibiotic as well. This issue will be discussed in more detail in the session on antibacterial resistance.

"To conclude my discourse, I wish to explain how we get the nourishment to carry out all these marvelous activities. Some of us, called autotrophs (self-feeders), can utilize inorganic sources of carbon, such as carbon dioxide gas, and derive energy from the sun or from the oxidation of inorganic chemicals. The rest of us (heterotrophs) must derive the energy to sustain life from the consumption of food, i.e., the oxidation of organic molecules. No, we do not possess a mouth or a digestive tract, as so wonderfully engineered among humans and other animals. Polysaccharides and protein macromolecules serve as our source of energy. Similar to humans snacking on a variety of foods, we secrete exoenzymes out into our immediate environment to hydrolyze available macromolecules into their smaller building blocks such as amino acids, mono- and disaccharides (sugars), etc. We possess a selective means for the diffusion and transport of these carbon- and nitrogen-containing compounds through our cell membranes. Once these substances reach our cytoplasm, our metabolic enzymes may further break the sugars and amino acids down into smaller byproducts, all the while releasing energy as this metabolic process called *catabolism* proceeds.

"Some of my colleagues have the capability to carry out catabolism in the absence of oxygen—these, humans call the 'anaerobes.' However, in the most common type of anaerobic catabolism (called fermentation), sugar molecules cannot be degraded totally, but rather stop in an intermediate stage. After coupling with hydroxide ions, these intermediate products end up as a variety of acid moieties which humans call 'mixed acids.' Because certain of these acids stink, many of my friends are, frankly, quite smelly. Humans refer to this as a 'barnyard' smell—which you might expect, as we thrive in the excreta of farm animals. Certain of my prokaryote friends also have the capability of releasing abundant gas from further breakdown of the mixed acids. You may have noticed from time to time, while attending this assembly, that a variety of gases and odors seem to permeate the atmosphere. This flatulence, if you will, is not obnoxious to us; however, humans with sensitive noses might find it offensive (but, cheers to our fellow prokaryotes that carry out these metabolic functions in sewage treatment facilities!).

"Others of my bacterial friends may perform this acid-forming feat in either the presence or the absence of air; humans call these versatile fellows 'facultative.' In fact, this is one of the major attributes by which you can tell our closely related cousins, *Staphylococcus*, from those of us in the breed of *Micrococcus*. Our *Staphylococcus* brethren can ferment glucose only to the stage of forming mixed acids, and/or gases as the case may be, and in the human lexicon are called 'fermenters.' However, my fellow *Micrococcus* buddies and I possess unique enzymes that allow us to extract considerably more energy than fermenters can, by breaking down carbon-containing compounds completely into carbon dioxide and water. This process is called aerobic respiration—even though we have no lungs—since oxygen is used as the final electron acceptor.

"A relatively few bacteria can use an inorganic electron acceptor in the absence of oxygen, in a process called anaerobic respiration, producing nitrogen gas, methane (swamp gas), or even poisonous hydrogen sulfide gas. Speaking of nitrogen gas, we should not overlook our friends the nitrogen-fixing bacteria, which are the only organisms on Earth able to break the triple bond of atmospheric nitrogen and generate the organic forms of nitrogen essential for the growth of all life."

Micrococcus luteus looked out over the assembly and observed that metabolic activity was declining to a low pitch. "Wake up, my good fellows! I realize that all of this may be familiar to you and seem a bit pedantic. But I am almost finished, and then you can hear marvels from *Enterobacter*, who is quite a more active and interesting fellow than I.

"You see, our narrow clan of bacteria do not engage in aggressive or alarming behavior. You can find us in various places, often on the skin of humans and animals. But we really are decent fellows. Humans use the term 'commensal' (a word that originally referred to those who 'eat at the same table' [*com mensa*] with others) for any prokaryotes that just hang around in a given niche without causing any harm. Such a designation fits us very well, as we produce no noxious substances, damaging enzymes, toxins, or anything else that may interfere with the normal activities of humans and animals. The ability of us micrococci to produce beautiful pastel yellow and orange pigments is perhaps our only claim to fame. How appropriate that this one attribute led to my name, *luteus*, a simple Latin word that refers to anything that has a yellow appearance. Some inquisitive and persistent researchers have determined by complex analyses that our pigment is a carotenoid. We just like how it looks.

"One last general comment before I turn the podium over to my colleague *Enterobacter aerogenes*. No matter what actions *Homo sapiens* take in an attempt to eradicate the prokaryotes, at least those species they consider of the greatest threat to them, we simply have too many tools at our disposal to resist annihilation. We replicate too rapidly, produce mutations at frightening rates, or acquire plasmids with DNA-encoded genes that often enhance rather than diminish our virulence and allow us to resist the actions of many antibiotics. This is all within the Darwinian scheme of existence. The great majority of our mutations either are nonfunctional or go to waste; however, it only takes one in a million mutations to pass the selection of species test. This surviving mutant then requires only a few hours to replicate into numbers that may produce a new and emerging disease or a newly resistant strain.

"Further, we can hide in too many sanctuaries, either in the form of resistant 'resting' spores, in a dormant state of reduced metabolism, or protected by proteins or other substances, in such a manner that we will never be eradicated. Humans must reconcile themselves to this reality. Not that they should—or will—give up their prime goal of containing or eliminating us, particularly when we produce life-threatening illnesses or mass infections in what they call 'outbreaks.' But humans should never operate under any grandiose illusions. We have not been nor ever will be totally annihilated. Given only a short time and the right circumstances, we will pop up again, either to repeat well-known infections or to emerge in a new set of clothes to cause yet-unimagined new forms of disease.

"I will hold for any questions until the end of *Enterobacter*'s discourse. I am sure others of you will also want to share several unique attributes that you have inherited with your genomes."

Micrococcus luteus rolled off the podium, making way for its colleague. With a brilliant wafting of its surrounding flagella, much like the coordinated efforts of oarsmen propelling a galley ship, *Enterobacter aerogenes* quickly found its place at the head of the assembly. The motion of its flagella came to a standstill as *Enterobacter* shut down the kinetic energy of its basal bodies and settled onto the podium. But as it looked out over the assembled horde of bacteria, it could see hints of fatigue here and there. In a flurry of renewed energy, *Enterobacter* again put its flagella in motion, unfolding much the way a peacock would spread its feathers, to draw the attention of its sluggish colleagues. The undulating flagella caught the light, refracting it in an array of rainbow colors like a living prism. The iridescent waves of light and sweep of wind from the waving flagella energized the assembly, bringing every cell back to a peak of attention.

Finally satisfied that the audience was now agreeably alert, *Enterobacter aerogenes* again let its flagella droop, like an array of flags falling limp on a windless day, to a more decorous position for its discourse.

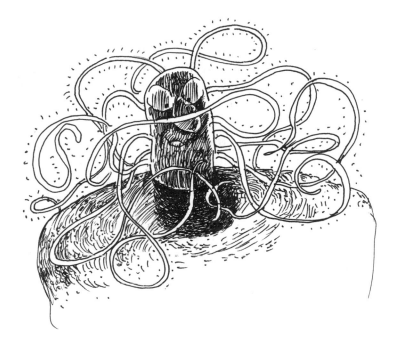

Discourse of *Enterobacter aerogenes*

"My fellow prokaryotes," bellowed *Enterobacter aerogenes* with resonating waves, "welcome to the continuation of this session. I can promise you what you are about to hear will not be dull.

"First, a word about these flagella that I have been waving so profusely and that so effectively caught your attention. It's interesting to compare our flagella with the mechanisms of locomotion of larger organisms: the flagella and cilia of the eukaryotic protozoa, the tentacles of jellyfish, even the wings of birds. The wings of a hummingbird beat an incredible 60 to 80 times per second, many times faster than other organisms, certainly much faster than most bacterial flagella. You will note I have many flagella myself, surrounding my entire outer cell membrane (humans call this 'peritrichous,' that is, 'hairs all around the body'). Yet all these mechanisms are designed to function with the most efficient use of energy, an amount unique to each species.

"Only those humans who have spent some time studying electron micrographs will appreciate just how intricate our flagella really are. Let me describe exactly how they work.

"First, each flagellum is anchored firmly to the cell membrane, extending beyond the cell wall at its outward end and reaching out into the surrounding environment. With my group of prokaryotes, that is, the gram-negative bacteria, the cell membrane 'anchor' is a series of rings that surround the rodlike base of the flagellum. The rings function not only as points of attachment, but also as bearings, if you will, that reduce friction and leakage of cell substance at the points of flagellar insertion. Each flagellum is only 13 to 17 nanometers (nm) in diameter (a nanometer is one one-billionth of a meter, or 1/1,000,000 of a millimeter; not even the breadth of a fleck of dust). The flagellum itself is a semirigid assembly of long, parallel, helical-shaped protein fibrils which are encased within a sleeve or 'hook.' The hook in turn is plugged into a 'basal body' consisting of a series of rings planted deep in the cell wall, with the inmost pair of rings located in our cytoplasm. And what makes it move? A nerve center within the basal body is activated by the movement of protons from the cytoplasm through the basal body rings, causing the flagellum to rotate like a propeller.

"How could something so infinitesimally small be so intricate? Humans must find this absolutely amazing.

"One more point, before I move on, that is even more amazing. The proteins that make up the flagellar fibrils have a different composition in virtually each strain of flagellated cell. Each differently composed flagellum, reflecting varying assemblies of amino acid sequences, is individually antigenic, so that antibodies are produced specific for each flagellar type. Humans refer to these antigens as 'H antigens' (the H from the German word *Hauch*, meaning 'breath'). Imagine, we are so fine and delicate that humans look on us as no more than a 'breath' of air.

"The result of all this H antigen business is that humans have found an additional way to subcategorize members of our clan, based on the specificity of antibodies that may be stimulated. Thus, the 'dog tag' for my notorious colleague *Escherichia coli*

O157:H7 can be translated into a cell wall or somatic antigen known as O157, with a flagellar antigen type H7, in contrast to other *Escherichia coli* strains that may have flagellar antigens of H1, H2, and so forth. All of this 'breath' business has assisted humans in singling out a given prokaryote tagged with the responsibility for causing infectious outbreaks or epidemics, much as a human in a police lineup would be identified by recognizing certain distinctive features.

"Let me now move forward to a more detailed description of our cell structure and how it differs from that of our gram-positive or 'gram-blue' friends. I would like to add here, by the way, that I see no reason to follow the human convention of referring to ourselves as gram *positive* and gram *negative*, as these are strictly human inventions. A gram-negative prokaryote has just as many positive qualities as a gram-positive one! These differences in staining qualities only mirror the basic differences in our cell wall structures, a part of our persona since the onset of time. It would have been more palatable from the beginning had humans learned to refer to us prokaryotes simply as 'gram blue' and 'gram red,' do you not agree? Thus I recommend this usage for the duration of this assembly, so that we may eliminate any insinuation that one bacterium is endowed with more or less positive or negative qualities."

A visible movement flowed among the motile bacteria in the audience in response to this suggestion, like the enthusiastic "Hear! Hear!" in the British House of Commons. *Enterobacter aerogenes* acknowledged their support with another ripple of its graceful flagella.

Just then, the pungent odor described by humans as a 'musty basement' began to permeate the air, and *Nocardia brasiliensis*, with its branching cells looking like forked tongues, was asking for recognition. "Very well," grunted *Enterobacter* with a burp of lactic acid, "what is your comment, *Nocardia*?"

"Indeed, I apologize for interrupting your discourse, and I shall be quick," replied the bacterium. "I ought to have spoken earlier, when the issue of Gram stains was first presented. We *Nocardia* possess one further unusual structure that relates to this gram-positive and gram-negative business. You see, we and our close cousins the mycobacteria are neither gram red nor gram blue. Our beautifully branched cells are coated on the outside with a waxy film, which effectively prevents the penetration of any water-based dye into our cell wall. Therefore, at least in the Gram stain technique, we remain untouched and are barely visible. Humans fumbled around for a long time until they discovered our sheltering waxy coat and how to detect us in spite of it.

"More times than not, humans seem to make new discoveries only by stumbling in the dark. The tale of how my cousin *Mycobacterium tuberculosis* was initially discovered by the now famous Dr. Robert Koch is a case in point that I hope you will find intriguing. Now, *Mycobacterium tuberculosis* is endowed with a waxy coat even thicker and more substantial than what we possess, so the Gram stain never reveals its presence. The good Dr. Koch knew that a bacterium was at the bottom of the dread disease tuberculosis, but had never isolated the organism. He went through scores of experiments, 270 to be exact, in which he tried every stain known to the scientific world at the time. He made up carefully measured solutions, one after the other, all to no avail. The Gram stain failed, as did a host of other water-based stains that had been successful in disclosing most of the other bacteria then known. Then one day, out of the blue (a bit of a pun, as that was how the mycobacterial cells looked with his methylene blue stain when observed under the microscope), he saw the slender, curved rods of the dreaded *Mycobacterium tuberculosis*. Perhaps his immediate response, 'Where, oh where have you been hiding?,' should be placed among the top ten best spontaneous quotes of the century!

"The astute Koch then scoured through all his notes trying to find out what was different about experiment 271, when all the rest had failed. No immediate clues. Suddenly he became aware of an acrid odor in the room. From some experiment conducted by his colleagues, the air had become saturated with ammonia. He then remembered that the staining solution of methylene blue he used that morning had been made the day before and had remained exposed all night to the air, as someone forgot to put the lid on the staining dish. Alkaline water! That was the secret—alkaline water.

"And so," continued *Nocardia brasiliensis*, "my story comes to a quick conclusion. Other scientists, Ziehl, Neelsen, and Kinyoun, among others, soon discovered that if our waxy coat were softened, dyes could easily permeate our outer mantle. Heating us up short of boiling, treating us with surface-active agents such as Turgitol, or soaking us in alkaline water were the methods used to soften our coats. Very powerful, acid-based dyes were used, and carbol fuchsin in particular was found to best permeate through the softened wax and paint our cell wall a brilliant red color.

"There is one last twist to the story. Once we have taken in the dye and our waxy cell wall is allowed to harden again, the acid dye becomes immovably locked inside our bodies. No amount of alcohol or acid washes can dislodge it. Thus, the mycobacteria, the nocardiae, and several other of our colleagues also with waxy coats will forever be referred to as 'acid fast.'

"I thank you for this opportunity."

"Thank you, *Nocardia*, for sharing this interesting story with us." *Enterobacter aerogenes* gave a dramatic wave of its flagella in appreciation. "It is indeed intriguing to learn of the unusual and ingenious characteristics we prokaryotes can boast, and as well to note how often pure serendipity has led humans to discoveries about us.

"Let us return now to the more basic question of our different cell structures, and how they relate to our behavior in the Gram stain.

"Indeed, *Micrococcus luteus* is to be commended for its wonderful analogies comparing the means and ways of our microworld with that of humans. The comparison of the structure of the gram-positive, or gram-blue, bacterial cell wall with that of the frame construction of a human dwelling is particularly masterful. Following through on that theme, the cell wall structure of us gram-negative, gram-red bacteria must be likened, in contrast, to that of a tent or teepee, with a thin, curtainlike inner lining.

"Our cell wall is much thinner than that of our gram-blue colleagues because we do not possess their thick, porous, sugary peptidoglycan coating. We do have a peptidoglycan component, but it is only one or a few units thick, with only one-dimensional cross linkages, forming a thin covering much like a porous plastic wrapper, not as dense or rigid as the gram-blue cell wall. Thus, the crystal violet dye so integral to the Gram stain, though it will stain us, is not trapped. The alcohol douche damages our outer membrane just enough to allow the alcohol to enter past the cell wall and quickly wash away the violet crystals. We become transparent again. To see us, scientists have to use that second dye or 'counterstain,' safranin. This gives us the pink or red color which humans simply refer to as gram negative.

"The simple fact that our cell walls are thin does not reflect our intriguing complexity. Indeed, our cell wall structure is perhaps even more detailed and interesting than that of our gram-blue counterparts. Outside our thin peptidoglycan layer, and anchored to it by a special lipoprotein, is a very complex outer membrane or shell composed of a mixture of polysaccharides and lipids, the latter component of which humans for some reason have called 'lipid A.' I will try to simplify my description of this outer membrane as much as possible; however, a visualization of the components is necessary to better understand what this membrane means to our well-being.

"The polysaccharide included within the lipid A layer consists of two unique sugars, an eight-carbon acid known in human abbreviation as 'KDO' (humans are too busy to bother with spelling out '2-keto-3-deoxyoctulonic acid') in association with a seven-carbon, heptose sugar. The chemical structure of this core polysaccharide is well conserved within a given genus, but may vary slightly between the species—which means in plain language that it pretty much is all the same from cell to cell. But, extending outward like whiskers from the core polysaccharide are what are known as 'O-specific side chains.' These side chains contain a galaxy of uncommon carbohydrates, providing unique antigenic structures by which our thousands and thousands of cousins and friends can be distinguished from one another. Our good friends in the family *Salmonella,* for example, have over 2,000 unique antigen types just within their clan alone. Just as no two humans are alike because of the infinite mixture of DNA base pairs within their individual gene pools, so we gram-red bacteria can be similarly individualized by the differences in the antigenic composition of our O-specific side chains.

"Now, back to our outer, lipid A layer. The lipid component is composed primarily of hydroxylated fatty acids, some of which are similar to fatty acids within human cells, but others of which are unique. It is the unique segments that can cause such misery for humans. In the process of a human infection with one of my gram-red friends, particularly when they get in the bloodstream, a condition known as bacteremia or septicemia may occur (the latter term meaning that humans react in some way to the presence of bacteria in the bloodstream). As lipid A is released into the bloodstream, either spontaneously or from dying cells, 'gram-negative shock' results. Humans experience varying degrees of fever, vascular collapse, shock, hemorrhage, and even death, depending on the amount of lipid A released and individual sensitivity to the uncommon fatty acids. Though I do not believe we can be held at fault if nature has endowed us with chemical structures that are neither possessed nor recognized by humans, resulting in such dire adverse reactions.

"I have additional wonders even more amazing to relate. Since our outer membrane is so thin, what allows us to absorb needed nutrients, while still blocking out toxins, antibiotics, and other undesirable molecules? Well, within the outer membrane are a cadre of unique proteins—logically known as outer membrane proteins. Among these are a unique set of porin proteins, the singular structures of which provide for transmembrane 'channels' through which most small molecules are allowed to transmigrate the outer membrane. In addition, we have specific channel proteins to aid the passage of specific molecules, including nucleotides and some sugars. Once in through the outer membrane, small uncharged and lipid-soluble molecules can pass on through the inner cytoplasmic membrane, which is semipermeable. Charged molecules or hydrophilic macromolecules must depend on special carrier molecules (such as permeases) or ion channels to cross the inner membrane.

"Now, I put the word 'channels' in quotes because these porins may not represent actual holes in the membrane—although some structural analysts believe they've identified 'trimer' molecules, identical proteins that indeed do provide a doughnut-like pore. The pores may be electronic in nature, with the porin proteins recognizing certain complementary chemical configurations or charges of the molecule to be passed, much as a human needs to enter a password or code to open an electronic lock. The particular configuration of each porin protein allows certain molecules to pass through while others are blocked out, particularly those substances that could be harmful to us. This mechanism is one means by which we can protect ourselves from the action of certain antibiotic agents, and we'll discuss this in detail in the session on antibiotic resistance.

"One last topic before we open the session for questions and comments. Humans have further subdivided us gram-red bacteria into groups depending on the way we metabolize and specifically obtain energy, either from carbohydrates or other sources. Their first concern is to identify those among us who are incapable of deriving energy from lactose: the so-called non-lactose fermenters. This subdivision is of importance to them because among the non-lactose fermenters are *Salmonella* and *Shigella*, two of our colleagues that have been charged with causing much human suffering in the form of diarrheal disease or dysentery. In fact, *Shigella* has been periodically so overwhelming that entire human armies have experienced enough of the 'trots' to be stopped in their tracks, so to speak. During the American Civil War, official records indicate that as many battles may have been decided from the ravages of our energetic relative *Shigella* as from gunfire. The manner in which *Shigella* accomplishes its mission will be presented in more detail in the next session.

"Many of you, besides, have undoubtedly heard of Typhoid Mary, the legendary cook who, refusing to wash her hands, spread *Salmonella typhi* far and wide to unsuspecting victims. Why should we bacteria be held responsible for causing so much misery when humans don't take care of themselves, practice abominable hygiene, eat out of one another's mess kits, don't wash their hands after eliminating their contaminated wastes, and so on and so on? But I will forgo that soapbox and get back on track.

"I will amplify only briefly on what *Micrococcus luteus* has already told us about fermentative and oxidative derivation of energy. Remember that fermenters are capable of degrading glucose and other carbohydrates only to pyruvic acid, a pivotal compound from which side reactions result in the production of a variety of mixed acids, alcohols, and even gas. In fact, humans capture certain of our prokaryote friends and enslave them in large vats to grind out acetic acid, formic acid, alcohols, and other products for their own consumption and profit. Occasionally, careless humans end up with the wrong species for a batch of grape juice, making a mess of the whole process. Then they complain bitterly and blame us for producing an unsellable wine that tastes more like vinegar, or even formic acid, eroding their profit margins. Serves them right!

"Each bacterial species has a profile of sugars that it can ferment. As mentioned before, those of us who cannot derive energy from lactose are called non-lactose fermenters. In like manner, there are non-sucrose fermenters, non-maltose fermenters, and so on. Others of our gram-red friends do not derive energy from sugars in this manner, but rather have the capability of marching forward from pyruvic acid into what is known as the tricarboxylic acid cycle (so called because most of the intermediary molecules contain three COOH, or carboxylic acid groups), first theorized by a man named Hans Adolph Krebs (you guessed it: this form of metabolism is better known among human circles as the Krebs cycle).

"The magic of this cycle is that, through a series of reactions involving the uptake of oxygen (the reason why this process is also called 'respiratory catabolism'), glucose and other sugars are degraded into the ultimate end products, carbon dioxide gas and water. Many times more energy is released this way than by fermenters, who leave much residual energy locked in the form of pyruvate and a variety of mixed acids. Then there is yet another group of prokaryotes that do not use sugars for energy at all, but pluck carbon and nitrogen from a variety of chemicals through an alternate set of enzyme-driven reactions. However, some of these 'asaccharolytic' bugs are a bit peculiar, and regrettably time does not permit further discussion of their quirks."

Having finished the formal presentation, *Enterobacter aerogenes* asked for questions and comments from the 'pond' (considering that ponds have bottoms and not floors). Looking out over the assembly, it noticed the thin, undulating form of *Leptospira biflexa* gracefully waving to and fro in an attempt to gain attention. "I believe our good spirochete friend has another contribution to the assembly," said *Enterobacter aerogenes*, making a motion with its flagella in the direction of *Leptospira*. "We're happy to hear what new information you have for us."

Spontaneous Effusion from the Spirochetes

"Thank you, *Enterobacter*, for allowing me to provide yet another brief interjection," replied *Leptospira biflexa*. "I applaud you for so eloquently describing your cell structure, particularly the microanatomy of your flagella. It is indeed marvelous how the flagellar structure, so tiny, can vary so widely. The dramatic way in which you approached the podium with your multitudes of flagella flying can only be described as stately and ceremonious. In contrast, note the graceful, almost musical rhythm that can best describe our sinuous motion. As I explained in the earlier session, we spirochetes also possess flagella, but far fewer; also, instead of flailing out in the air, ours are almost completely internal, attached to and encased within a tough outer sheath. Thus, with our characteristic spiral shape, we gracefully rotate around this helical axis, advancing gently the way a serpent or eel slices through the water.

"Indeed, I am one of the spirochetes, the name with which we were tagged by an early human investigator who presumably thought that I and several closely related colleagues looked like a wisp (*spira*, Latin for 'coil') of long, flowing hair (Greek *chaitē*) when he peered at us through his primitive microscope. As regards frequent name changes, we can also sympathize with *Moraxella lacunata* and *Acinetobacter,* as finding our place in the human lexicon has not been without frustration. Actually, we never acquired the formal name of *Spirocheta*. The closest we came was at one juncture when we were called *Spironema*, with the suffix *nema* meaning 'thread'; or, later, *Microspironema*, because someone thought we were quite tiny. Now let me introduce myself as what my name is today—*Leptospira*. *Lepto*, of course, means 'thin,' so that pretty well describes what we and most of our spirochete cousins look like: thin and coiled up like a corkscrew.

"But this is not the reason for my coming before you in this session. In this discussion of staining and other visualization methods, we spirochetes and our cousins can provide yet another interesting example. Indeed, we are not very visible at all when humans look for us under the microscope, and special techniques are needed to even see us. Our dietary habits have kept us thin as a rail; in fact, if a human were as thin per length as we are, he would appear no more than 2 inches wide. Then the name *Homo spironemus* might be appropriate! But for an image of sheer beauty, I would invite each of you to observe us in a dark-field preparation, a clever human invention for observing minuscule, poorly staining, moving microbes. They submerge us in a drop of water and place us in a microscopic field illuminated by slanting rays of light that bounce off our bodies. There we are—bright, fluid, undulating threads of light against a black background when viewed through the oculars, much as shooting comets might be observed against the dark of a moonless sky. I specifically would invite you to observe a company of a thousand of us as we perform our undulating dance, there in the fluid mount of aspirated material from an infected human.

"By the way, it may interest you all to know that many years ago we leptospires signed a housing agreement with rats and other animals to provide a base from which to operate. So we and our animal friends now live harmoniously together—with one peculiarity: our narrow corkscrew shape and gentle motility allow us to wriggle easily through the animal kidney, from whence we are passed out into the soil and water systems. Then along come human veterinarians, animal lovers, water sports enthusiasts, and workers in sewers, abattoirs, rice fields, and other environments frequented by our intimate friends the rats, cats, dogs, and the like. By one means or another, these humans ingest urine-contaminated soil and/or water.

"Once in our new home in the human body, we make a beeline for the blood, taking side trips into the small blood vessels everywhere in the body. This unfortunately tends to plug the works wherever we end up. You can trace this activity by observing the blotches, hemorrhages, and rashes we produce all over the victim's skin. In some cases, it may be difficult to tell our effects from those of the Black Plague. By the time we reach the liver and kidney, our aggression is maximal, and these organs begin to fail. The resulting condition observed in humans undergoing our invasion carries the almost unpronounceable name of 'icterohemorrhagica,' describing the yellow skin (*ictero*, from the Greek for 'jaundice') covered all over with red blotches.

"And in this regard, good friend *Enterobacter*," *Leptospira biflexa* finished, "I would request that you also allow a brief postscript from my distant cousin *Borrelia*, who has recently gained notoriety by making the top 10 list of emerging human pathogens. I believe you all have heard of Lyme disease."

Borrelia took the flurry from *Enterobacter*'s flagella to be a sign of approval.

"In contrast to *Leptospira*," began *Borrelia* with an air of confidence, "our pact is with ticks and lice, but still, we inadvertently cause human misery. These insects are opportunistic creatures which humans, particularly when camping and hiking in the wild, or perhaps resting in unsavory and unsanitary environments, have little chance of avoiding. When a tick steals a blood meal from our host, we come along for the ride, much the way our colleague *Yersinia* does with its flea carrier. The gut of the louse and the salivary glands of the tick offer convenient reservoirs in which we can hole up. Then, when the louse bites an unsuspecting human, we are released under the skin, just like *Yersinia*.

"Once inside the human, our threadlike cells distribute widely through the circulation, causing skin rashes similar to those described by *Leptospira*, accompanied by fever, chills, muscle aches, and confusion. The threadlike cells can become dormant, holed up in macrophages or deep within the lymph nodes. Symptoms subside, and those infected humans believe we have left the scene. But then we reemerge sometime later in an instant replay. Humans descriptively call this 'relapsing fever.' Only recently gaining notoriety, *Borrelia burgdorferi* in particular likes to take up residence in the membranes of the joints, producing a new chronic arthritic disease called Lyme disease, named after Old Lyme, Connecticut, the town in which the first recognized outbreak occurred. Our very successful cohabitation with the Eastern deer tick has made this disease one of the most common arthropod-transmitted infections. Indeed, bonny little Bambi may not necessarily be as sweet and harmless as he appears!"

"Thank you, *Leptospira* and *Borrelia*," replied *Enterobacter aerogenes*. "I am pleased that you have shared these most interesting adaptations with us. Now, fellow prokaryotes, our time is nearly finished. We thank you for your attendance and trust that you have gained insight into our physical being, *modus operandi*, and certain interactions with other creatures and humans as well. You should be left with a good feeling that, in much of our structure and physiognomy, we prokaryotes are as intricate and complex as eukaryotic cells. We shall now prepare for the next session, to be conducted in a panel format."

Session 3:
Microbial Pathogenesis and Human Infection

Moderator:
Stenotrophomonas maltophilia

Panelists:
*Staphylococcus aureus,
Pseudomonas aeruginosa,
Escherichia coli*

Contagium animatum: It is not the contagium, but the disease itself which is a parasitic organism, or, more ambiguously, is a parasitic life process. No! It is not the disease, but the cause of the disease, which reproduces itself.

—J. Henle, 1840

Thermotoga maritima

The panel format itself may have predisposed to the high interest in this session, contributing to the packed house, or, in this case, to the packed pond. After all, it is not often that notables such as Staphylococcus aureus, Pseudomonas aeruginosa, *and* Escherichia coli *assemble at the same time. This trio, who together cause the*

The Other End of the Microscope

majority of bacterial infections, are among the most famous and interesting of the prokaryota, drawing attention much the same way as the Three Musketeers, the three fateful Norns of mythology, or the three witches on the heath of Macbeth's castle do within human circles.

Although exchange of genetic information between bacterial cells has always been widespread, occurring spontaneously through conjugation (the approximation of two cells from which nucleic acid segments can be transferred from donor to recipient), exposure to new and broad information has been relatively limited for each clan and generally confined to relatively closely related species. Thus, the information offered in this session on the ways by which bacteria become aggressive, and the mechanisms by which interactions with humans produce deleterious effects, generated considerable enthusiasm. Even we bacteria have our "learning curves."

Among the panelist celebrities, Staphylococcus aureus *may appear the most regal, with its golden yellow color catching the rays of the afternoon sun and simulating the halo (aureole) seen around the heads of saints in medieval paintings. On close observation, the beautifully regular, spherical cells in their grapelike clusters could not have been more perfectly arranged in the vineyards of a formal medieval garden.* Pseudomonas aeruginosa *also presents a striking pose, surrounded by an eerie green color. Each individual cell stands tall, a single flagellum protruding from its top like a feather or plume from the top of a hat. In comparison,* Escherichia coli *looks a bit shabby, certainly not playing the role of the great celebrity. Its individual cells are short and squat, dull gray on the surface, and covered with delicate, hairlike fibrils. It seems to have no desire to conform to any particular style or image. Nonetheless, of the three, this one which appears the most unimpressive can relate an array of virulence mechanisms that, in the end, will elicit the most amazement and agitation among those assembled. Perhaps it is a general rule of nature that one can never judge the nature of any living "monad" based on its external appearance.*

Stenotrophomonas maltophilia *was selected as the moderator of the session. You may wonder, reader, why a bacterium with such a long, unpronounceable name would be selected as moderator. Again, the name is purely of human invention; prokaryotes in general have no particular craving to assume a title. In fact, the name* Stenotrophomonas *was the last of three attempts made by humans. Its initial appellation was* Pseudomonas, *which experienced morphologists found was not plausible because it did not look or "quack" like other pseudomonads. Just for the record, our colleague was quite happy to escape this unseemly name, as* pseudo *("impostor" or "hypocrite") and* monas *("unit" or "speck") give the impression its clan is little more than a "fake crumb," not a flattering nickname.* Xanthomonas *was the next tag, based on the yellow pigment of its mature colony. The bacterial geneticists threw this one out because the* Xanthomonas *genes simply were not there, forcing the taxonomists to scurry for yet another name.*

The name Stenotrophomonas *most likely was the result of a committee deliberation. As this species is somewhat narrow or con-*

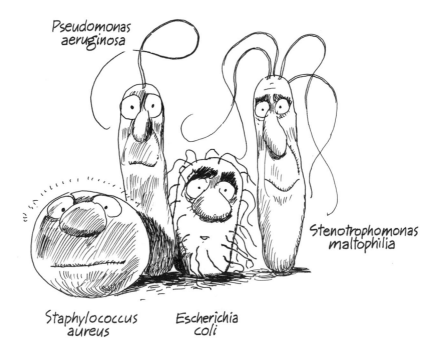

Pseudomonas
aeruginosa

Stenotrophomonas
maltophilia

Staphylococcus
aureus

Escherichia
coli

The Other End of the Microscope

strained by the nutrients it can assimilate and metabolize, some bright scholar contributed the word steno *("restricted"), also used as the stem for the word "stenosis." Then, adding the stem* tropho *("to grow") and retaining the suffix* monad, *humans invented a new word,* Stenotrophomonas. *To achieve some balance,* maltophilia *was selected as the name extension, as this organism has an insatiable craving for the sugar maltose. Thus the name for the bacterium that has problems growing except in certain milieus, particularly in environments deficient in maltose.*

Stenotrophomonas *has lately found its revenge for this nomenclatural insecurity. By occupying several niches in the hospital environment and by shuffling its genes, it has evolved antibiotic-resistant strains which cause nosocomial infections in debilitated humans. More of this will be related presently by its closely related cousin,* Pseudomonas aeruginosa.

Stenotrophomonas maltophilia's Introduction

Stenotrophomonas maltophilia mounted to the podium twig and addressed the gathering: "Welcome!" Although spoken with a firm and purposeful intonation, its greeting barely penetrated the buzz and hum among those assembled. "Welcome," *Stenotrophomonas* repeated, "to this assembly. I will not take long, as I know you are most anxious to hear from each of those comprising our famous panel.

"But harken first to a few words that will paint a background to our relationship with human beings. Our interaction is usually described by the term 'infection,' which can be somewhat misunderstood. *Infectus* means to 'dye,' 'stain,' or 'contaminate.' Infection, as applied to bacteria, merely indicates our penetration into human tissues, whether or not a disease process is present. In most cases, following penetration, we merely go about our business and basically go unnoticed. In this state, we are often referred to as being 'commensals,' which, as has been mentioned, means to habitually 'eat at the table' together.

"Now, some of us prokaryotes have the capacity to thrive in tissues, replicate rapidly, and cause disease. In this state, we are referred to as pathogens, or as being pathogenic; that is, we are among those agents that cause *pathos* or 'suffering.' Some have greater or lesser capacity to cause suffering; or, as humans say, are more or less virulent—*virulencia* being a Latin word that means 'extreme bitterness' or 'malignity of temper.' Those of us who are more venomous are referred to as being of 'high virulence,' in contrast to our more temperate, 'low-virulence' friends. This word has also trickled down to refer to those particles of nature even smaller than we, namely the viruses, though it seems to me to be a bit redundant to speak of 'virulent viruses.'

"The main focus of this session is to learn of the several mechanisms by which bacteria exercise virulence, which humans refer to as virulence factors. The ability to express virulence is bred in our chromosomes, or embedded in the plasmids that we acquire in our journey through life. In fact, humans are now able to fracture our chromosomes, exposing the secrets of the three-letter codes (codons) in our DNA, as explained in the last session by our colleague *Micrococcus luteus*. The more industrious of the human scientists have actually sketched out our codons and are discovering that certain molecular sequences are related to specific virulence factors.

"The virulence stories that each of our panelists will relate are nothing less than fascinating. For example, you will hear from *Escherichia coli* about its production of a toxin that causes watery diarrhea. This toxin, for all intents and purposes, is virtually identical to cholera toxin and similar to Shiga toxin. The production of this toxin is governed by the same DNA code in the chromosome of each of these bacteria. In the future, it may not be necessary for laboratory microbiologists to identify *Escherichia coli* versus *Vibrio cholerae*, for example, as the specific agent of watery diarrhea. Rather, recognition of the nucleic acid sequence encoding the toxin that produces diarrhea ('breaking the code') will establish the disease process, and treatment will be to administer the appropriate antitoxin.

"Some scientists predict that certain human nucleic acid sequences may code for traits that predispose their bearers to be thieves, murderers, adulterers, or other 'virulent' conditions. Will there be arrests if a chromosome sequence presaging an adverse character trait is discovered in a genetic study, even before a crime is committed? Will human fetuses be aborted if chromosomal analysis of amniotic fluid cells reveals potential psychological predispositions? Such *a priori* 'cleansing' perhaps is not beyond the realm of twisted contemplation by certain humans.

"But enough of this philosophical rambling. Each of our three panelists will relate modes of contact and penetration of humans and their mechanisms of virulence and production of pathogenic side effects. Time will be reserved between presentations and at the conclusion of this session for others of you here assembled to describe virulence factors unique to your clan, perfected by natural selection over billions of years. Now I present our 'golden boy,' *Staphylococcus aureus.*"

Discourse of *Staphylococcus aureus*

"Fellow prokaryotes," *Staphylococcus aureus* glowed. "I am pleased to be with you to share certain of the mechanisms by which we cause problems for humans and animals, and how we protect ourselves from annihilation once we gain access to their tissues. Our predisposition to cause disease, which humans call virulence, is through the production of a variety of enzymes and toxins. Perhaps the most important product we *aureus* staphylococci manufacture is coagulase, an enzyme that causes blood to clot, resulting in a fibrin webbing such as *Yersinia pestis* has already mentioned. From the action of hyaluronidase (an enzyme that melts connective tissue), lipases, and other proteolytic enzymes, we can carve a space within the substrate of tissues and hollow out a cavity in this meshwork. Then coagulase again goes to work, laying down an outer fibrin framework around this space, much as a weaver would fashion a wicker basket. 'Abscess' is what humans call this protective handiwork. We perhaps can be called the basketweavers of the prokaryote empire.

"This retaining wall serves to protect us from assaults by our predators. During any given tissue invasion, we lose a host of our members; a major portion may be simply eliminated through the action of white cells or phagocytes that swarm to every site we attempt to occupy. The battle is fierce within the abscess space, as humans readily appreciate when they squeeze the thick yellow pus from the center of a boil or furuncle. We do have other pieces of armor that help to protect us from the wave after wave of phagocytes. Through evolution, we have accumulated in our cell walls and outer capsules a group of key proteins as a defense both against the phagocytic or ingesting prowess of the white cell scavengers and against the human immune response. In particular, leukocidin, by punching holes in the membranes of phagocytic leukocytes, provides us with the capability to escape the destructive digestive action of these scavenger white cells. Once this mechanism is in place, these hungry monsters simply cannot touch us.

"Further, a special molecule incorporated in our peptidoglycan cell wall, teichoic acid, allows us to stick to plastics, particularly as may be used in prosthetic devices. Teichoic acid–rich strains can easily set up housekeeping on intravascular catheters, often leading to devastating local infections and 'blood poisoning.' My close associate *Staphylococcus epidermidis* is uniquely endowed with teichoic acids in the form of 'slime factor,' a sticky substance that not only allows it a firm grasp on the surface of prosthetic devices, but also offers protection from the ever-present cellular and immune antibacterial forces.

"This piling up of organism after organism on a plastic surface, within the abundance of polysaccharide slime, produces a thick mass referred to as a glycocalyx or biofilm, the closest that we prokaryotes have come to evolving into a multicelled 'organ.' In order for the bacterial cells at the bottom of the slime to survive, a mechanism is necessary to deliver nutrients and remove waste products. So, we organize. Channels are opened through the maze

BIOFILM COMMUNITY

of slime factor, facilitating the delivery of vital nutrients through all layers of the biofilm community. These same channels are shaped and reshaped to export waste products from the bottommost layers. Thus, a 'breathing,' 'feeding,' 'excreting' biofilm community progresses along the course of the plastic object.

"This cooperative effort also provides protection against adverse environmental changes and neutralizes the action of toxic agents in our immediate ecosphere. We are so well organized within this community that even high doses of antibiotics have difficulty in removing us from the scene. Once we gain a foothold on an intravascular catheter, the only option to prevent a blood-borne infection is for humans to remove the catheter, along with our biofilm community.

"Outside of these biofilm communes, a perfectly healthy animal will ultimately win this 'battle of the abscess.' But give us a slight opening, such as defects in the chemical signals that draw white cells to the site of an infection (humans call this 'chemotaxis')—or mount a less than dedicated immunological response by the serum activators called opsonins—and we may well win the battle. Perhaps little known to most humans, we staphylococci have also developed an intricate system of signals by which we detect changes in the environment of the abscess and can make a defensive response. A discussion of this signaling mechanism, intricately governed by the activation of gene-controlled sensor and response regulator proteins, is a fascinating story, but beyond the scope for presentation at this assembly.

"If we hang around in an abscess long enough, we can gain access to the lymphatic channels, through which we can spread widely in the skin, cropping up in multiple sites in the form of a condition humans call 'impetigo.' Or, if we gain access to the bloodstream, scooting along with our adhesin pili flying high, we can stick to tissue receptors, forming metastatic abscesses in virtually any organ. Some of our clan have developed mechanisms to resist the action of human-improvised antibiotics, the mechanisms of which will be described by a colleague later in this session.

"Our major mode of causing human disease is through the production of toxins. Now the word 'toxin' has an interesting origin, with equally fascinating implications. *Toxicon* is the Greek word for 'bow poison,' referring to the venom into which warriors dipped their arrows. The picture of a poisonous arrow leaving the string of an archer's tightly pulled bow is a perfect metaphor for the release of a molecule that has its effect on cells at some distance. One toxin that humans call 'exfoliatin' is a prime example. The target organ is the surface cells of the skin, the epidermis. Exfoliatin causes a shedding of that outer layer of skin, resulting in what humans call the 'scalded skin syndrome.' We also produce a number of cell membrane–damaging toxins, tagged with Greek letter prefixes (alpha toxin, beta toxin, etc.), that have as target organs the destruction of red cells, white cells, and even human tissue and organ cells, referred to as somatic cells (*soma* = 'body').

"Most notorious of our recent onslaughts on *Homo sapiens* is toxic shock syndrome, caused by a potent toxin, TSST-1, that was acquired by certain nasty members of my clan probably through the acquisition of a plasmid. Most of you know the original story. Some women used hyperabsorbent 'super' vaginal tampons during menstruation to allow them to assume normal activities for prolonged periods of time. It was not long before we discovered that a tampon left in place for prolonged periods played beautifully into our hands. We first of all had a perfect place to hole up. The outflow of secretions was stopped, preventing any chance to 'wash' us away, and we were able to establish substantial colonies. But of most danger to the unsuspecting women was the amount of iron immediately available to us. Unknown to humans until just recently, the presence of iron in our milieu not only stimulates maximum cell reproduction, but increases both the amount and the potency of TSST-1 that we can produce. Menstrual blood contains plenty of iron ions, in addition to the carbon- and nitrogen-rich red cell degradation products which promote our replication and growth even more. The inserted tampon introduces oxygen into the normally anaerobic vaginal canal, further encouraging the production of toxin. So, many of these poor women had no chance—our TSST-1 opened up the walls of the blood vessels throughout the body, causing a general release of fluids into the tissues which led to low blood pressure, fainting, shock, and ultimate demise from heart and kidney failure. It was only when the Denver pediatrician Dr. James Todd realized the beautiful environment these high-absorbency tampons were creating for us that human women were warned about their dangers.

"On the more innocuous side, you will recognize our effects of getting 'under the skin,' so to speak, in the form of the familiar furuncle and carbuncle. It may be a different story if we get a hold in traumatic wounds or surgical incisions. We reside in so many niches, and can hide in so many crevices in the skin, that we pose an ever-present threat. Even surgeons, after multiple hand washings, may not totally eliminate us from tiny skin crevices. Wherever we can invade into human organs, we set up abscesses, causing much morbidity and even death if the human happens to be debilitated.

"I have talked much too long and will now turn the podium over to my colleague *Pseudomonas aeruginosa*. In closing, I would like to remind humans, particularly those who demonize us for what we do to them, that we are a legitimate part of creation and have every right to pursue our existence. Create atmospheres or niches where we can thrive, and you must bear the responsibility for all the havoc we can cause. This is fair warning! Wash your hands frequently, keep your food refrigerated, take proper precautions when creating or sewing up skin wounds, and don't let blood accumulate anywhere in our proximity. We have no preconceived plan to harm humans; rather, we simply carry out the activities prescribed by the genes in our chromosome. Put a group of fraternity brothers around a keg of beer, and you know what happens. When conditions allow us to marshal all of our potentials at the same time, new diseases such as described will continue to emerge."

Spontaneous Effusions

Staphylococcus aureus rolled off the podium, moving to the area on the rocky crag reserved for the panelists. *Stenotrophomonas maltophilia* glided up in its place, inviting questions or comments from the assembly.

Streptococcus pyogenes asked for attention. Receiving an inviting bend of recognition from *Stenotrophomonas*'s plumelike flagellum, *Streptococcus* began to speak.

"I do not want to unnecessarily prolong these proceedings," it apologized. "However, I do have additional virulence factors to describe to the assembly. I will represent others of my *Streptococcus* clan, of which there are many members, divided by humans into 'beta-hemolytic' and 'alpha-hemolytic' groups, depending on the way we attack and affect the red cell membranes.

"First of all, most of the enzymes we produce work differently from those of *Staphylococcus aureus*. Admittedly, certain of our clan can also produce a TSST-1–like toxin. Could there be similarities between us and *Staphylococcus aureus* that are more than cell wall deep? It may be that a future human scientist may find similar nucleic acid sequences in our respective genomes, giving genetic validity to a common ancestry. But that aside, we more commonly produce proteolytic enzymes, primarily one humans call 'streptokinase.' Instead of effecting fibrin deposition and abscess formation, streptokinase dissolves clots and opens up passageways between the tissue fascial planes. As a snowplow clears the way after a storm, allowing traffic to pass once again unimpeded, so our enzymes open up passages through which we spread quickly into the tissues. Instead of being confined to abscesses, we roam freely, producing a diffuse reddening of the infected part, called 'cellulitis.' Humans also refer to this as 'erysipelas' when the skin and subcutaneous tissue are our primary target. You all have heard of what humans call 'the flesh-eating bacteria'? Yep, that enemy is us.

"Now this erysipelas is different in the condition humans call 'scarlet fever.' This rash is produced by the production of a toxin—an erythrogenic toxin, which attacks the walls of blood capillaries. Cells lining these vessels loosen up, allowing plasma and red cells to pour through into the extracellular tissue. The affected skin becomes swollen and beet red in appearance, particularly the face, neck, and chest and soft-skin areas of the arms.

"We also have one other important phenomenon. As you know, the 'sore throat syndrome,' particularly in children, technically called 'acute pharyngitis' in scientific circles and 'strep throat' among lay people, is also our doing. The local invasion of the throat tissue is bad enough, causing reddening, pus formation, and pain. This all is short-lived, however, and rarely produces permanent damage—but here is where the plot thickens. Lower- and higher-order animals that possess a lymphoid cell system—including humans—produce protein substances called 'antibodies' in response to any invading molecules considered to be different from those of the host, or 'non-self' if you will. These antibodies act specifically to counteract or neutralize offending proteins (known as 'antigens'). We do not escape this reaction, and, in fact, our cell bodies can be ultimately eliminated by the action of specific antibodies.

"But here is the intrigue. Antibodies produced against antigens in our cell walls cross-react with similar antigens present in other host tissues, at least in humans. What does this mean? Doctors theorize that these cross-reacting antibodies attach to streptococcal-like receptors on the synovial membranes of the patient's joints, the inner endothelium of the heart and kidneys, and similar tissues of other organs. In a few weeks after the sore throat subsides, a delayed antigen-antibody reaction ensues at these sites, resulting in secondary inflammation of the heart, joints, kidneys, and other organs, depending on the target. The patient suffers rheumatic fever, debilitating diseases of the joints (synovitis and arthritis), and inflammation of the heart (endocarditis) and/or the kidneys (nephritis), or what is called the 'poststreptococcal syndrome.' Thus, in the end, even though we may have been eliminated from our settlement in the mucous membranes in the throat, we still have a delayed effect.

"For the metabolic peace of all you prokaryotes, particularly those of you who also may be caught in the middle of a natural phenomenon that is out of your control, we feel we bear no primary responsibility for these secondary events that can cause debilitating effects in humans. The fact that nature created joint, heart, and kidney membranes with structures similar to our cell bodies is simply an example of gene synchronicity, a random error phenomenon that is totally beyond our control."

"Thank you," responded *Stenotrophomonas maltophilia*. "Do we have further comments?"

"*Streptococcus pneumoniae*, here. I have one further comment. First of all, I want to thank our close cousin *Streptococcus pyogenes* for disclaiming responsibility for natural phenomena that are out of our control. As you know, I and my immediate clan are responsible for a special type of pneumonia that can be rapidly fatal in some people, particularly babies, the old, and those who have allowed the ravages of smoking and drink to damage their bodies. Indeed, the illness of close to half a million residents of the United States per year has been laid at our doorstep, and from 5 to over 30% of these unfortunate people die. What must the toll be worldwide? This is not to mention that we also, regrettably, are responsible for several hundred cases of fatal meningitis a year and about half the cases of middle-ear infections in children.

"Innocently—from our perspective, anyway—our genes code for the production of one piece of protective armor: a thick polysaccharide capsule. Just as a human puts on a thick raincoat to protect him from the elements, so our capsule protects us from the onslaught of the armies of white cells that try to eat us alive, a process that humans call 'phagocytosis.' They simply cannot touch us. Protected by our capsules, we continue to proliferate, invade the lung or tissue covering the brain, and cause considerable inflammation. It is not until several days later, when anticapsular antibodies finally reach such a level that they whisk away our protective coats, that the white cells can finally swallow us whole. Then the infection subsides. Unfortunately for the patient, in many cases such delay is too late."

Streptococcus pneumoniae
(with protective coat)

"Thank you so much, *Streptococcus pneumoniae*, for sharing that most interesting aspect of your physiology," replied *Stenotrophomonas maltophilia*. "In the interest of time, I must move forward and introduce *Pseudomonas aeruginosa*, who has yet other intriguing mechanisms of virulence to share with us."

Waving gently back and forth, *Psudomonas aeruginosa* wriggled skyward to the top of the projecting twig, pulling itself up with its long, apical flagellum. As it commenced its oration, a distinct grapelike aroma could be detected in its immediate surroundings, which grew increasingly noticeable as the bacterium's energetic presentation continued.

Oration of *Pseudomonas aeruginosa*

"My fellow prokaryotes," began *Pseudomonas*. "I shall briefly get my personal irritation out of the way at the onset, before modulating to something more substantive. As has been mentioned before, the demeaning nature of our name *Pseudomonas*, which translates into something equivalent to a 'fake fleck,' has already been revealed. Who among you would accept such a name for your christening!? At least the name extension *aeruginosa* is in a more positive vein. *Aeruginosa* is derived from a Latin stem meaning 'replete with copper rust,' in recognition of our unique ability to produce a verdigris-green pigment. As this is a pleasant pigment to view, we bear no grudge for this part of our name.

"It has been estimated that only 0.1 to 1% of all microorganisms in the global ecosphere have been identified and scientifically named. With the advanced technology currently possessed by humans, literally hundreds of these hidden microbes will be discovered. We hereby humbly make the plea that, if humans insist on giving any newly discovered microbe a name, it should carry a positive spin, heralding the attributes of the species involved rather than focusing on minor deficiencies as has been the practice in the past. We also request that humans reserve their family names for business firms or products of human invention and not saddle us with a label that may be completely unconnected with our proud heritage. Having so vented my metabolic vapors, I will let this matter rest.

"One can only marvel at the tremendous diversity that exists among us, the prokaryotes. Through our rapid replication and short doubling times, we have great capacity to adapt via mutation to virtually any environmental adversity. Even though as many as 10^6 or 10^7 of any given population of prokaryotes may be lost to some force, particularly antibiotic action, even one cell in a million that survives thanks to a unique self-sustaining mutation can soon replicate into a resistant clone. This assessment should provide us with great optimism: that, even as minute as we are, we collectively possess such a reserve of basic nature that our long-term survival is virtually guaranteed.

"This matter of diversity leads me to share with you one of our outstanding attributes. As already briefly mentioned by *Pseudomonas fluorescens*, our clan can survive and replicate at temperatures as high as 45°C. This property may seem to be of little moment; however, it has broadened our ecosystem and has positioned us to be of greater worry to our human friends. Look at our history. Through the ages we have essentially been content to live in warm aquatic estuaries all over the world. In truth, we are an extremely aquaphilic (water-loving) bunch. If prokaryotes were to stage Olympic games, we would repeatedly score high among the top swimmers. Similar to the story related by *Legionella pneumophila*, changes in human habits have, through no fault of our own, placed us in new positions where interactions with humans have been greatly accelerated. Constitutionally we are a mild-mannered race. We at best are known as opportunistic pathogens, which means that we rarely pose a problem to normal, healthy people. We become bothersome only when we are provided with a moist or wet atmosphere and an opportunity to invade through an open port into a host who is debilitated or immunosuppressed.

"Now, look at the recent history of humans in relation to the bacteria. First of all, humans as a race are living longer. A high percentage of elderly people become debilitated, with a lowered immune status. Patients with chronic illnesses are admitted to hospitals where water environments much to our liking have been created. The elevated temperatures of certain of these water reservoirs provide us with an optimum environment to reproduce at accelerated rates. Additionally, many tubs and nebulizers are not cleaned often enough, or with effective enough levels of antiseptics, to hold us at bay. Couple this with the increased number of surgical procedures where moisture often aggregates at sites of barrier breaks. One of my clan has a good chance of infecting every tracheostomy site, every burn, every indwelling catheter, or any site where moisture accumulates. This is not because we have any innate desire to do so. You might just as well put a cow into a green pasture and tell it not to eat.

"As is the case with *Legionella*, we pseudomonads have not changed our living habits enough to account for the worldwide increase in bacterial infections. Rather, changes in human behavior and practices, opening up favorable ecosystems in which we can survive and maximally multiply, must be held responsible.

"Oh yes, we pseudomonads are not totally innocuous. We can swim around quite freely via our single flagellum. Being able to oxidize glucose all the way to CO_2 and water (respiratory), we possess maximal efficiency to sustain our activities. We simply do not wear out. We possess a well-developed set of delicate fimbriae and pili that allow firm anchor to the cells of humans and animals. This property of attachment, called adhesion, which *Escherichia coli* will mention in greater detail, is necessary for us to set up housekeeping at any given site of potential infection. And lastly, we produce a wide array of enzymes, many of them cytotoxic and of a proteolytic nature, providing us with the ability not only to destroy cells but to liquefy everything in the immediate surroundings.

"This latter property comes into crucial focus if we ever gain access to a damaged eye. The liquor produced in tears is particularly to our liking. Humans have learned, often through tragic experience, to marshal every mode of defense whenever there is any hint that we've taken up housekeeping in the conjunctival sac, and have also learned to prevent our access to the bloodstream at all costs. What they call '*Pseudomonas* septicemia' is, in some environments, fatal in about two-thirds of all those who are so infected. We fit very well the adage, 'Beware the fury of a patient being.'

"And so, my fellow prokaryotes, this is my story. Thank you for the opportunity to be with you."

Stenotrophomonas maltophilia, realizing that the day was moving on, indicated that only one more comment would be allowed during this juncture.

"*Burkholderia pseudomallei*, here. First of all, we wish you to know, *Pseudomonas aeruginosa*, that we empathize with the recent taxonomic upheaval within your clan, in which so many of your former brethren, including ourselves, have been delegated by humans to other geno-groups, leaving your family most sparsely populated. This is not with our total approval. As you indicated in your opening remarks, we were not unhappy to shed the trivial name of *Pseudomonas*; however, I am not sure that gaining the new designation *Burkholderia*, named after a decent and still-living human investigator, has been an improvement for our identity. My reason for speaking, however, is to mention one other ecosystem in which we, as former members of your tribe, have found a home—the estuaries and irrigation trenches in southeast Asia. We caused considerable misery among recruits in the Vietnam War: 'trench foot' is what they called it. The feet of soldiers, subject to constant moisture either in boots or barefoot, provided us the opportunity to invade. By a quirk of nature, we also have the capacity to survive in the tissues for months and even years. In recognition of this feat (no pun intended), we have acquired the interesting appellation, 'Vietnam time bomb,' ready to reactivate and explode any time the local environment becomes inviting."

"*Burkholderia cepacia*, here," came a quick addition from the same sector of the pond, close on the heels of its cousin before *Stenotrophomonas maltophilia* had the opportunity to intercede. "I would like to mention one other habitat where those of our clan have caused a problem, mostly involving economics—just to let you and the assembly know that we *Burkholderia* not only invade and cause havoc in humans, particularly those with cystic fibrosis, but in animals and plants as well. My extended name, *cepacea*, is based on the genus name of an onion, *caepa*. Onion rot is my specialty. I also speak on behalf of *Burkholderia gladiolii*, the name extension of which indicates that it can affect other bulbous plants as well, specifically the roots of many domestic flowering plants."

"Thank you for your comments," quickly interrupted *Stenotrophomonas*. "As we have been in session for some time, let us have a short intermission. I invite the audience to take this opportunity to blow off a bit of gas or exercise your flagella, as the case may be."

During this break, *Pseudomonas aeruginosa* slithered off the podium, opening up space for *Escherichia coli*. This exchange of positions was somewhat artistic, reminiscent of a tribal dance, with the long flagellar plume of *Pseudomonas aeruginosa* undulating in broad, sweeping motions, in contrast to the rhythmic ascending beats of the shorter, peritrichous flagella of *Escherichia coli*. This provided those in the assembly with a bit of entertainment during the interlude.

"And now," continued *Stenotrophomonas maltophilia*, "I have great pleasure in introducing our final panelist for this session, *Escherichia coli*. In the realm of pathogenesis and virulence, this distinguished colleague probably possesses more mechanisms for virulence and antibiotic defense than any other prokaryote. It is with great expectation that we await a full disclosure of its activities. *Escherichia coli*, the assembly is yours."

Oration of *Escherichia coli*

"Thank you for the warm welcome, *Stenotrophomonas*," began the modest-looking bacterium. "Indeed, as I look at myself and my immediate clan, I can only gaze with wonderment. This is not any attempt to puff myself up in self-worth. Rather, every living creature must see itself as a wonder of nature. Even we one-celled prokaryotes, so tiny that many thousands of us can occupy the space taken by the period on a printed page, are infinitely complex, as already alluded to by other presenters in this assembly.

"The capabilities that I and my closely related bacterial colleagues possess, the functions and activities that we are capable of carrying out, and the good and evil that we can accomplish, on which I will elaborate in a moment, are simply astounding. In truth, we did not directly inherit all of the current activities of which we are capable. Granted, our chromosome includes the genes and nucleic acid codes necessary for us to maintain our basic metabolism and to produce the many products characteristic of our species. However, most of the virulence factors that I will be describing to you momentarily have been acquired through the activities of millions and millions of our progenitors. The basic genetic codes to achieve most of these functions have been gained through the assimilation of plasmids acquired from some of our closely related friends.

"How do we acquire these plasmids, you may ask? This process is probably the closest that we bacteria come to animal or human 'sex.' Humans call our process 'conjugation,' based on the Latin term *conjugalis,* or 'united in marriage.' Such an analogy stretches the point a bit for us, except that in this form of genomic transfer, two bacterial cells must come into approximation. This differs from other forms of gene transfer in which we merely imbibe bits of extraneous chromosomal or DNA material released from dying donor cells (transformation), or when similar bits of genetic material carried by a viral phage penetrate through our cell wall (transduction)."

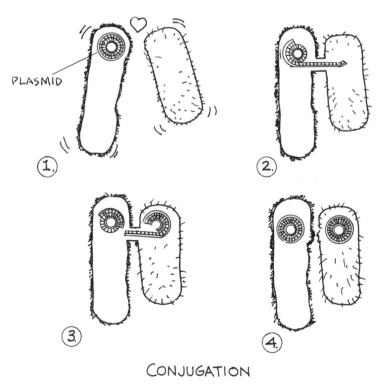

CONJUGATION

"Conjugation is very different. Not every cell has the capability to conjugate. This unique skill is limited to those cells that have acquired a particular plasmid called, for some unknown reason, the 'F' plasmid; in human terms, these specially enabled cells are called 'F+'. I myself am F+, and grateful for it. Now, what does this allow me to do? If I can get close enough to any of you other gram-red bacteria, whether a fellow *Escherichia* or any of you from many other clans, I can grab hold tightly to your cell wall, extend my F-induced pilus into your protoplasm, and inject key portions of my gene material. Thus, I may be able to transfer certain traits that would provide you with new metabolic prowess—or, to the consternation of humans, I could pass on to you various resistance factors that I have acquired through several generations. You now would also be able to withstand the same forms of natural or man-made antibiotics. And key to this process is that you would also inherit the ability to pass these newly acquired traits on to your progeny. Or, as humans would say, 'another resistant mutant has emerged.'

"For example, at one time many generations past, one of my *Escherichia coli* ancestors received via conjugation from our friend *Shigella dysenteriae* a plasmid that allows us to produce a shigella-like toxin. (In truth, the shigellae are so closely related to us that it was probably a human oversight that we were ever separated in their taxonomic scheme in the first place.) This Shiga-like toxin mediates intestinal invasion and cell destruction in humans. Micropockets of inflammation occur, where millions of bacteria invade loci on the intestinal mucosa, resulting in painful, severe, bloody dysentery. These effects are identical to those resulting from *Shigella dysenteriae* infection; in fact, we cause 'shigellosis.' Those of us who have this plasmid are called 'enteroinvasive *Escherichia coli*.'

"Although we *Escherichia coli* are considered as a single entity in terms of basic phenotypic properties—that is, in the way we look and in the way we metabolize—not all members of our clan are necessarily so closely related. Unique differences exist between us, much as there are wide variations within subgroups of *Homo sapiens*. First of all, each of our cells is surrounded by tiny, almost submicroscopic hairlike extensions known as fimbriae. These fimbriae are not structurally identical from one of my cousin species to another, any more than human hair is identical. In fact, it is specifically these differences in fimbriae by means of which some members of our *Escherichia coli* clan carry out functions that others within our same clan are incapable of. It is through the action of the fimbriae that we may or may not be able to attach to a given animal or human cell. Without attachment, we cannot colonize, and without colonization we lie loose and are simply washed away, incapable of causing either good or evil.

"Compatibilities and incompatibilities exist within each bacterial family, even as they do between every creature. Humans don't let just anyone into their house; nor do they associate or 'conjugate' equally with every other human. Why should we be different? Therefore, only certain *Escherichia coli* strains can conjugate with a *Shigella* that may possess a trait advantageous to us. Others of us will conjugate with fimbria-compatible bacteria of other species, each carrying unique plasmid codes that have resulted in great diversity within our family.

"Based on these fundamental differences, humans have divided us into groups based on certain acquired virulence factors— enterotoxigenic, enteropathogenic, enteroinvasive, enteroaggregative, enterohemorrhagic, uropathogenic, and so forth. These terms mean little to us—fundamentally, certain of our members simply act differently because we have acquired different plasmid-encoded genes. Thus, some of us demonstrate virulence traits like *Shigella*, others produce effects similar to *Vibrio cholerae*, others like *Salmonella*, and so forth. And why is the way we act similar to these colleagues? Simply because we have acquired these traits from them through conjugations somewhere in our ancestral history.

"I have discussed my *enteroinvasive* cousins and how they work, damaging the human intestinal mucosa to cause severe dysentery. Now another group of us, called the *enterotoxigenic* strains of *Escherichia coli*, a bit more mild-mannered, also present a most interesting set of circumstances. What I am about to explain is virtually identical, with minor exceptions, to the manner in which *Vibrio cholerae* causes watery diarrhea, in this case produced in the absence of local destruction or inflammation in the intestinal mucosa.

"The *Escherichia coli* enterotoxin, structurally similar to the cholera toxin, is composed of two subunits. Subunit B is responsible for the binding of the toxin to the intestinal epithelial cell membrane. Subunit A in turn possesses two peptides: A_2 facilitates penetration through the membrane; once inside the cytoplasm, the toxin peptide A_1 separates from A_2 and is activated.

"Perhaps the best way I can explain the action of this toxin is to compare it to the invasion of an imaginary valley by an alien spaceship. The five landing pads beneath the spaceship represent the subunit B, or binding portion of the toxin. These five pads allow the ship to land only on a specific landing site with complementary receptors, similar to the surface of an intestinal epithelial cell. Within the spaceship are the aliens, or subunit A, some of whom carry chisels (representing the A_2 peptide subunits) and others carry wrenches (the A_1 peptide subunits). Now, down the center of our imaginary valley runs a river that is evenly dispersed into irrigation channels throughout the valley. The water in those channels is retained there by valves which are enclosed behind a tall security fence (comparable to the intestinal cell membrane) near the spaceship landing site.

"The aliens with the chisels (A_2 peptide subunits) leave the spaceship and open a hole in the security fence. The aliens with the wrenches (A_1 peptide subunits), close on their heels, head straight for the valves and crank them open so that the irrigation water drains out of the channels and flows straight down the main stream. The effect, if you will, is that the valley suffers from an acute surge of watery diarrhea!

"Now, let me describe what actually happens within the intestinal cell. I realize this will get a bit detailed, but crank up your ribosomes momentarily because this is interesting. Once the active toxin-carrying peptide A_1 subunit enters the cytoplasm—

breaks in through the security fence, as it were—it carries a message for a stimulatory protein to activate a compound known as adenylate cyclase. The function of adenylate cyclase (comparable to the wrenches carried by our A_1 aliens) is to activate cyclic AMP (cAMP), and cAMP in its turn pumps chloride, bicarbonate, and potassium ions *out of* the intestinal epithelial cells *into* the bowel lumen. It also blocks reabsorption of sodium out of the lumen. There is an old adage, 'Where the ions go, there goes the water.' All those chloride, bicarbonate, potassium, and sodium ions in the lumen have the powerful effect to pull water, through osmosis, out of the vascular system and interstitial tissue, flooding the lumen. The result is repeated bouts of copious watery diarrhea, much like the uncontrolled flood of water gushing down our hypothetical river.

"Unfortunately for the victim, there is a kind of snowball effect in this situation. The pH of the diarrheal efflux, rich in sodium, bicarbonate, and chloride ions, is very high. A high pH is one of the major stimuli for the growth of *Vibrio cholerae*—which is why alkaline peptone water is used for its selective recovery from stool specimens in laboratory cultures. The more organisms, the more A_1 toxin is produced; more A_1 toxin raises the alkaline ion concentration; and the higher the pH, the higher the concentration of bacteria.

"So, my good colleagues, this is the action of *Vibrio cholerae*, mimicked by some strains of my clan. Thus the enterotoxigenic *Escherichia coli* can serve as an alternate agent of human 'cholera.'

"Now I will describe yet another mechanism by which certain of our *Escherichia coli* family cause human suffering. The *enterohemorrhagic* subgroup (actually a very small group, which humans have tagged with the designation O157:H7) has gone out of control, in my opinion, in its assault on the human race. Too many children have succumbed to their distemper. Indeed, we quite disapprove of these black sheep cousins of ours, just as humans condemn others whose purposes are cruel or destructive. Yet, the story of the O157:H7 enterohemorrhagic subversives, though a sad tale, is interesting in its own light.

"The toxins produced by the O157:H7 clan were also acquired from *Shigella*—in fact, humans refer to these villains as the STEC strains, that is, the Shiga-toxin *Escherichia coli* strains. Though we can't blame *Shigella* for what happened to the genes they shared; even humans can be duped into innocently causing great ill. Anyway, the Shiga-derived toxin, as conjugated by the O157:H7 strain of *Escherichia coli*, turned out to be terribly potent. The human intestinal mucosa is seriously damaged, resulting in bloody diarrhea of varying degrees of severity. The diarrhea itself usually resolves in a few days, but an urgent complication will affect about 5% of all humans who acquire the infection: the 'hemolytic uremic syndrome.' Toxin is adsorbed into the bloodstream and directly alters or destroys the red blood cells. Clotting factors are released, causing the microvasculature to shut down, much as was described by *Yersinia pestis* in its tale of the Black Death. The kidneys and other organ systems fail, too frequently resulting in rapid death.

"For reasons unknown, the O157:H7 villains have a great predilection for colonizing the intestines of dairy and beef cattle. Thus, contaminated ground beef, particularly uncooked or undercooked, has caused most of the diarrheal and hemorrhagic shock outbreaks. Transmission of infections via raw milk or even well water contaminated with cattle manure has occurred. Care must be taken in the ingestion of undercooked sausage, unpasteurized apple cider, and raw vegetables, which also may be contaminated. Transmission between humans can also take place, because only a small number of organisms can cause an infection and preventive measures such as hand washing are poorly practiced in some circles.

"The point of this grim tale is that a sequence of events transpired to produce a new type of disease. First of all, *Escherichia coli* is far more prevalent in the ecosystem than *Shigella*; therefore, the virulent genes that were transferred had far greater dissemination. The O157:H7 strain also possesses the complementary attachment machinery necessary to invade the human intestinal mucosa. With its preference for colonizing the bovine intestine, a secondary and effective means for transfer of organisms to humans evolved. Meats and meat products are carelessly processed in many parts of the world. Watch out for undercooked hamburger! Steak tartare should be absolutely on the forbidden list. Children are particularly susceptible to the secondary effects of the Shiga toxin. Humans, please beware!

"In closing, I almost failed to mention perhaps our most common human interaction. We are involved in urinary tract infections, by a mechanism not dissimilar to that which I have described for enteric infections. Certain of our members have acquired fimbriae or pili complementary to the receptors on the epithelial cells of the urinary tract. Humans call these the *uropathogenic* strains. Not only do these strains possess the necessary complementary antigens to effect adhesion, but they also produce other virulence factors that promote urinary tract infection, including hemolysins and protective anticomplement enzymes.

"We have also just learned of mutant strains of *Escherichia coli* in other ecosystems that produce attachment epitopes that are actually injected into epithelial cells. With this 'attachment protein' now internalized on the other side of the epithelial cell, infections are not limited strictly to the uropathogenic strains. Attachment is now guaranteed for virtually any strain! As these mutants become more and more prevalent, humans are faced with an expanded problem in attempting to contain urinary tract infections.

"There are other functions and virulence activities that could be presented, either unique to or shared by many of my *Escherichia coli* colleagues, but, although interesting, they are less common and beyond the time allotted here for presentation. Thank you all for your attendance and attention."

Spontaneous Effusions

Escherichia coli slithered off the tip of the podium, and *Stenotrophomonas maltophilia* again assumed its role as moderator of the assembly. "Good colleagues," it announced, "please remain for just a few more moments. I had promised three of our important toxin-producing friends that they might briefly and informally share their stories at the conclusion of this session."

Looking into the center of the assembly, where those gathered were in closest proximity to reduce the amount of ambient oxygen, *Stenotrophomonas* recognized *Clostridium botulinum*.

"Greetings from the world of the anaerobes," wheezed *Clostridium botulinum*, maintaining a steady outpouring of vapors to minimize any back-diffusion of oxygen, an activity that produced a bit of a stench in the immediate environment. "Indeed, the story I have to relate also has special interest, not only to those here assembled, but to any humans who may review these transcripts.

"Where do you normally find us clostridia? Well, we prefer to live in the soil, but will subsist in any environment in which only bits of contaminating dirt may be present. You many also find us deep in the muck beneath lakes and seas, where oxygen is limited. You see, we have a particular tendency to produce spores whenever environmental conditions are a bit marginal. This is a property that we share with our distinguished colleague *Bacillus anthracis*. Our only difference is that we produce spores in the absence of oxygen—*Bacillus anthracis* needs plenty of air. We have the common attributes that our spores can survive virtually forever and can resist virtually every environmental change, including boiling water.

"What difference does this make? Visualize a good Hausfrau, preparing canned preserves or vegetables for consumption during those cold winter nights. Perhaps her canning jars have not been properly cleaned. Even though she boils the jars just before adding the preserves, any bits of residual dirt provide our spores with just enough space to survive. We welcome the addition of the hot sweet preserves or nourishing vegetables just before the cap is tightly affixed to the jar. Not only do we have more food to attack than a cow in a 10-acre pasture, but the lack of air is just what we need to germinate, particularly if the food is not acid and the jars are not stored at a cold temperature.

"Again through no fault of our own, we possess genetic machinery that codes for the production of a substance that happens to be extremely toxic to humans. The target of attack for our toxin is the tips of peripheral nerves, where the transmitter of messages, acetylcholine, is stopped in its tracks. No message transmission means no nerve action; and if no nerve action, then no muscle action, no heartbeat, no breathing. Kaput! In fact, some nasty-minded humans have spread the word that one small vial of 'botulinum toxin,' if dispersed evenly in popular canned products, 'can annihilate an entire city'—although they don't say how many people live in the city! If our action were all in one place, for example one city, it would not be long before we were discovered. However, because commercial canned foods are distributed widely throughout the world, small, widely scattered outbreaks could go unnoticed, unless a concerted effort at intercommunication is established. We can literally strike before folks know what hit them.

"One last idiosyncrasy that makes us even more interesting. Don't ask me why, but we really take after honey as a habitat. Perhaps it is the high concentration of sugar and the low concentration of oxygen. In any event, honey also happens to be one delectable that mothers like to feed infants—'a little bit of sugar makes the medicine go down!' The infant intestine, having a low level of acid and not being colonized with so many other bacterial species that would hinder our multiplication, provides an ideal environment to perform our tasks. The mother first notes that her baby will not suck, may start gagging, and may become inert and unresponsive, just lying there like a limp balloon. In fact, if we can get a sufficient hold, the baby may even die—which is not to our advantage either. Feeding honey to babies has been discouraged in some of the more informed and sophisticated circles. We approve; we are just as happy to stay in the honey jar and not cause any grief.

"Thank you for listening to my story."

"Ah yes," chimed in *Clostridium tetani*, who was positioned immediately adjacent to its closely related cousin, "our toxin also acts on the nervous system, but a bit differently. We don't live in honey, nor do we cause any effects to humans after being swallowed. Rather, our spores prefer to reside on rusty nails or the sharp prongs of barbed wire. These spores survive unchanged until that nail accidentally pierces the foot or hand of an unsuspecting human. If the sharp barb penetrates beneath the skin, the oxygen level in the deep tissue is so low we can 'hatch' (germinate) and go about our business. Many humans do not realize our ability to remain patiently in spore form for many months, even after we have gained access to their tissues. We are content to live in the depths of a wound for months on end, until some other disruptive force comes about.

"The tetanus toxin that we release also attaches to nerve endings, particularly those that stimulate muscle action. But instead of remaining at the tip of the nerve, this 'tetanotoxin' actually penetrates into the nerve and is distributed to active sites throughout the central nervous system. The effect is not to block transmission of nerve messages, as with botulinum toxin, but actually to block the release of those chemicals that allow the muscle to *relax* after a contraction. Thus, the muscles tend to stay in a state of prolonged contraction. Humans have invented the clever term 'lockjaw' when the muscles of mastication are involved, or 'risus sardonicus' when the facial muscles are in varying stages of grimace and contortion. Even we tend also to grimace when we witness the effects on the long muscles—the arching of the spine, the contorted flexing of the arms or hyperextension of the legs, and all the while the afflicted human crying out with pain and exhaustion. It is terrible that these natural byproducts of our existence cause such a tragedy for humans and animals. The widespread use of antitoxins and vaccinations has mercifully resulted in these extreme cases being less and less frequent."

"Thank you," commented *Stenotrophomonas maltophilia.* "My friends, allow me now to bring this session to a close . . ."

But just then, *Corynebacterium diphtheriae* rose to interject one last comment. "May I add one further point? I will be brief," it hurried. "We *Corynebacterium diphtheriae* also are to some extent a victim of circumstance. We normally inhabit the oral cavity of humans, living in peace with the world, not causing any problems. But if we are infected with an extraneous bacterial virus, called a 'beta phage,' we too transform completely out of character, just as a rabies-infected dog goes absolutely crazy. This phage is nasty in that it possesses the exact genetic material (the *tox* gene) necessary to unlock our carefully controlled propensity to produce diphtheria toxin when iron is limited in availability.

"This toxin first of all destroys many of the mucosal cells around our position in the human oral cavity, so that a thick, inflammatory pseudomembrane begins to coat the throat and deeper airways, cutting off the air flow. More importantly, this toxin also acts to block protein synthesis in distant cells, such as in the heart and respiratory nerves. Respiration and circulation come to a stop. Once this phage enters our cell, we are doomed to perpetual virulence and can cross over to other humans through an unprotected cough or sneeze. Unfortunately, in our maddened state, we find it easier to find a foothold in children, who are least prepared to withstand our destruction. Once we are phage possessed, we spread quickly from person to person and simply cannot control our actions. Insane humans were once thought to be possessed by demons; would that we could be relieved of our pitiful state by a divine exorcism of this phage! To date, this has not happened. We can only urge that human laws be enacted for mandatory vaccination."

"Thanks to all of you for such gripping stories," concluded *Stenotrophomonas*. "The time has arrived to conclude this session, so that this great assembly can move on to other business."

Just as the next session was about to be announced, *Mycobacterium gordonae*, its golden yellow countenance shining in the slanting rays of the sun, agitated for recognition.

"Fellow prokaryotes," it emitted in a metabolic burst felt throughout the immediate surroundings. "Chairbug *Stenotrophomonas*, please hear me out. We have one item of business to which we should attend. The stories of three more of our prokaryote colleagues, notables who may have caused more human misery and discomfort from the beginnings of measured time than any of the rest of us, must be heard. Unfortunately these colleagues are not able to join this assembly here, as their niches in the universe are confined to human habitation, and Prokaryote Pond has no trace of human 'contamination.' Even if humans had deposited my colleagues here, these surroundings would not sustain their viability, even for a short period.

"I thus request, *Stenotrophomonas*, that we extend this session just a bit longer. Why are we suddenly so obsessed with the urgency of schedule? When have we prokaryotes ever bothered about time before? Our 'lag,' 'log,' and 'stationary' periods of growth are only a human definition. Could we somehow, perhaps by osmosis, have acquired the human trait of impatience? So I ask that you grant me a few moments, my fellow prokaryotes. Experiencing the stories of my immediate cousin, *Mycobacterium tuberculosis*, and those of *Neisseria gonorrhoeae* and *Treponema pallidum* is essential for your total understanding of our ongoing face-off with humans."

While remaining at the rostrum, *Stenotrophomonas maltophilia*, looking over the assembly and noting that none present made any objection, offered a somewhat reluctant reply, "Thank you, *Mycobacterium gordonae*. Your words are wise, and indeed it is unfortunate that we almost forgot the important contributions made by our bacterial colleagues you mentioned. Please relate these stories as quickly as possible, as we still have one long session to complete before the Taxonomy Committee convenes."

"Thank you, *Stenotrophomonas*," replied *Mycobacterium gordonae* as it ascended to the podium. "Although each story about to be related could occupy an entire session, I will see that the comments are limited only to important bug-human interactions. Each of these stories is somewhat clouded in history, and certain facts have not been adequately transmitted from generation to generation, either through human ancestry or through conjugations of our own genomes. Therefore, much is left for speculation.

"One unresolved issue, which will remain obscure even after the conclusion of this session, is how and when did each of these bacteria evolve? If their niches are limited to humans, each could not have existed, at least in its present form, in a natural state prior to the appearance of humans on earth. One can only assume that each of these species gradually mutated from a rather different ancestor to appear relatively late in time. Through natural selection, and after multiple trial mutations, the precise chromosomal codes have evolved within each of these bacteria to permit a unique cohabitation with humans—certainly more to their disadvantage than ours, I might add—that has remarkably persisted over centuries.

"As each of these distinctive organisms has thus become withdrawn, like a human hermit, to a limited habitat in the tissues and on the mucous membranes of humans, they are not here to relate their stories directly. I have acquired plasmid-encoded messages from each, however, which I shall interpret for you. Let me begin first with my immediate cousin, *Mycobacterium tuberculosis*." *Mycobacterium gordonae* pulled out from under its waxy coat a long strand of messenger RNA and translated the following.

Message of *Mycobacterium tuberculosis*

"Greetings to you, fellow bacteria. My name is *Mycobacterium tuberculosis*, and I live distant from you in the byways of human tissues. It is unfortunate that I cannot assemble here with you, but I need a complex medium to survive and would not be able to last for long in the environment of Prokaryote Pond. In some ways such a limited choice of habitation is like dwelling in a prison; however, each living form on earth must choose the arena in which it can thrive best.

"What's in a name? My extended name *tuberculosis* is derived from 'tubercle,' from the Latin *tuberculum* or 'tuber,' referring to 'a small, knobby prominence or excrescence.' This description fits the appearance of the soft, white nodules that develop under the skin and in deep organs, due to the inflammatory reaction produced by the host at each site of colony formation. In a dramatic pamphlet published in 1903, Ernest Poole, a New York City journalist, dubbed tuberculosis the 'White Plague' to fix in people's minds that, even though the disease was much slower moving than the virulent Black Plague of medieval days, which spread like wildfire to leave thousands of victims dead in its wake, it was just as horrible and deadly. Tuberculosis has also been variously known as 'consumption,' 'phthisis' (the wasting disease), and even, most imaginatively, the 'Captain of the Men of Death' (by John Bunyan).

"In any case, our family name, *Mycobacterium* (*myco*, or 'fungus'), is simply derived from the observations of certain early scientists that these tubercles in victims' lungs closely resembled those observed in fungal infections. And so, the name *Mycobacterium tuberculosis* is now held in awe and fear by humans, rising from an age when as many as one in seven deaths in many places resulted from our patient, progressive attack.

"The association of our clan with humans extends back many centuries in time. Evidence is irrefutable, from examination of the skeletons of Neolithic and pre-Columbian humans, that our work had already begun ages ago. Although the entire clinical picture cannot be grasped from the examination of a few lifeless bones, I can assure you that these ancient people suffered the full extent of our effects. But during these early periods in time, and for several centuries thereafter, our activities were limited to only a few humans, widely scattered in time and space and largely out of communication with one another.

"However, when the human race in its 19th and 20th centuries entered into what their historians call the Industrial Revolution, we came out of hiding. People began to gather together in the great cities of the world, packed together in slums and poorly ventilated structures of all descriptions. We transfer from one human to another by clinging to the airborne droplets expelled when one coughs or sneezes into the face of another. We are very small fellows. Look under the microscope at our short, thin, curved cells: perfect for wiggling down the air passages to the depths of the lungs, all the way to the air sacs where the conditions are perfect to set up a new colony.

"Our ability to persist in this process through centuries of time can be attributed to two key characteristics: we grow very slowly, and we have been endowed with a thick, waxy layer serving as an outer coat. Our waxy cell wall (along with other factors) may help us to avoid destruction by the scavenger alveolar macrophages which come on the scene soon after we arrive. Oh, they gobble us up, but we live on right in their cytoplasm, our waxy coat protecting us from the destructive enzymes released during their lysozymal bursts. In fact, these macrophages become our allies, serving as a convenient taxi service to carry us not only to other areas in the lung, but ultimately to distant organs as well, where we can develop many new colonies, a pioneering outreach that is referred to in human circles as the 'miliary' form of the disease.

"With our slow growth, our human hosts do not experience a sudden illness; therefore, early on in the disease process, we are often ignored. But as our miliary granulomas develop into larger inflammatory tubercles, more and more of the host energy is diverted to these fields of action, to the extent that the rest of the body simply begins to waste away. The ancient Greek physicians used the term *phthisis* to describe this progressive wasting process.

"Until recently, no known treatment or cure was available, beyond the creation of several fresh air–type sanatoria where clean air, rest, good food, and pleasant conversation were the only antidotes to our gradual progress. But recently we have been faced with the onslaught of powerful drugs, often used in combination, which for a time put us back on our heels. In fact, we came so close at one juncture to total ruin that humans began to circulate word that we tubercle bacilli were fast on the road to extinction. But it did not take us long to catch on. As our replication is so slow, many of the drugs only put us in a temporary state of dysfunction, but we still remained alive. Unseen by our human hosts, we initiated a slow process of mutation, resulting in the emergence of progeny that progressively became resistant to one drug after another.

"Two external events moved this evolution even more rapidly in our favor. Because the periods of treatment are so long, many humans either did not receive the necessary dosage of the drugs they were taking, or they did not have the patience to complete a full course of therapy. Each of these failings allowed our chromosomal engineers to assemble the necessary resistance codes to develop mutants with diversionary metabolic pathways that render the drugs ineffective.

"Secondly, we received a real boost from the human immuno-deficiency virus, which humans call HIV. That ally deprived an ever-increasing number of HIV-positive humans of their innate mechanisms of resistance to our activities. Not only were our drug-resistant progeny able to withstand the effect of antituber-cular drugs, but they also shortened their doubling time; that is, they replicated more rapidly than the common strains. After that short period of elation, humans are again faced with the grim prospect that our new drug-resistant descendants may bring human tuberculosis back full circle to where sanatorium life will again be the only 'cure' that can be offered. To what extent this account will bring elation to you prokaryotes here assembled is a matter of individual discretion; however, at least you now know that we as a family have the resources available to protect our-selves and survive through otherwise dire circumstances. Thank you for reviewing my story."

Mycobacterium gordonae then consulted the next plasmid, paused for a moment, and again addressed the assembly.

"The following account," it resumed, "is provided to you by *Neisseria gonorrhoeae*, another of our colleagues that has had a cen-turies-old association with humans. To some extent, its tale is even more intriguing than what I just shared from *Mycobacterium tuber-culosis* because of its association with the intimate parts of the human anatomy, with sexual intercourse being the only ferry to get them from one person to another. Oh yes, some of you may have been informed that infection may result from contact with a con-taminated toilet seat, but this remains a virtual impossibility unless two individuals are simultaneously occupying the same seat! *Neisseria gonorrhoeae* cannot survive on a toilet seat long enough to retain its infectivity. Thus, a human infected with our illustrious colleague must carry the stigma that he or she has engaged in sexual relations, which may be a social as well as a medical problem if the contact person was, ahem, outside the appropriate relationship.

"Here is what *Neisseria gonorrhoeae*, hidden away in millions of urethral and cervical outposts throughout the world, wishes to share with you."

Message of *Neisseria gonorrhoeae*

"Good day to you, my fellow prokaryotes. For many centuries, the malady we cause, gonorrhea, and another closely related entity, syphilis, were considered to be the same disease. However, when it was discovered experimentally some time ago that material from a gonorrhea pox could not be induced to produce the chancre of syphilis when injected from one human to another, it became obvious that the two diseases were distinct.

"Our physical distinction from the threadlike syphilis microbes has been known since we were discovered in a Gram-stain preparation by the German microbiologist Albert Neisser, whose name we carry to this day. The way we neisseriae look has been ridiculed in some circles. We consist of nothing more than a pair of spheres, staining red in the Gram stain, lying in close proximity, with our apposing edges flattened. Some say we look like a couple of opposing coffee beans; others like the butt end of a monkey—not a very dignified image!

"Humans have chronicled when they believe we evolved, at least based on ancient writings describing the unique disease we cause. Theodor Roseburg, in *Microbes and Morals*, cites early Egyptian papyrus medical writings in which the need for urethral and vaginal injections to cure 'burning discharges' is mentioned. Historians also have also found reference in Chinese writings ascribed to Hoang Ty, as early as the third millennium BC, in which unique symptoms suggesting gonorrhea were described: 'affectations of the urethra and vagina, with drainage of corrupt materials white or red' (again as cited by Roseburg). A Hebrew account in Leviticus (15:2) also clearly seems to indicate gonorrhea—'When any man hath a running issue out of his flesh, because of his issue he is unclean.' Descriptions of 'running' diseases were common in both ancient Greek and Roman literature. In 150 AD, the famous Greek physician of the day, Claudius Galenus, known today as Galen, clearly recognized the purulent penile discharge. He incorrectly mistook it for semen, but the name he coined, 'gonorrhea' ('flow of seed'), has stayed with us.

"One of the more interesting 'name the disease' games played by humans is the term 'clap' applied to gonorrhea. It is our observation that humans apparently have no more respect in applying derogatory names to their own kind than to us microbes. Can you imagine how an afflicted person must feel, already bearing the shame of a sex-linked disease, then to have it publicly announced that he/she has the clap? It certainly became evident early on that this term did not have anything to do with applause. The term 'clap' appeared in English literature sometime in the 16th century. In the Middle Ages, a section of Paris was known as *Le Clapier*, which colloquially refers to a 'rabbit burrow.' This neighborhood included many houses of prostitution, and as a result the common French name for brothel was '*clapise.*' From that, the English truncation 'clap' was used to describe the 'running disease' acquired by visitors to any such establishment. Currently 'the clap' and 'the drip' are used synonymously, neither of which terms particularly dignifies the human race!

"We will let you in now on the secrets of exactly how we do our work. What one does not see on cursory examination of our double spherical cells are the very delicate, spider web–like pili which extrude from our outer membrane, making us look something like a fuzzy ball. These special protein strands have molecular arrangements that are complementary to structures (microvilli) on the surface of non-ciliated columnar epithelial cells, particularly those lining the human urethra and genital tract. This provides us with an organ of attachment. Like an insect's foot sticking to fly paper, or the recent human invention Velcro, with its hooks and fuzz, we stick tight.

"Once attached, we tell the epithelial cell to take us inside by a process called endocytosis. We then travel across the cell inside a vacuole and pop out the other side, by exocytosis, out into the subepithelial tissue. There we begin to multiply, inducing an inflammatory response (mediated by tumor necrosis factor) that leads to the destruction of the surrounding epithelial cells; this opens up a direct route for us to invade the soft, moist submucosal tissue. Soon a horde of scavenger white blood cells flood the scene, many of which successfully make a meal of us—or so they think. But once inside the cells, our pili again come to the rescue, warding off the destructive effects of the powerful intracellular lysozymal enzymes. In fact, we survive within the protection of these cells quite well. Thus, when humans observe the battlefield in a stained smear of discharge through the oculars of a microscope, there we are, waving back at them from *inside* the scavenger cells—'intracellular gram-red diplococci' is the verdict—now you have your diagnosis.

"The battle does not end there. We are so hardy that we can remain in human tissues for years, gradually expanding our occupation zones and riding to other parts of the genital system within the protection of our white cell taxicabs. The smoldering inflammatory reaction finally results in the laying down of a fibrin mesh, closing the various ducts that we inhabit. Thus, men cannot pee and women become sterile because their eggs become trapped in their journey down the Fallopian tubes. And so, full circle back to the Egyptian papyri and a better understanding of why ancient physicians were using injections and catheters to relieve the blockage and the burning.

"Because we have such a confined niche in nature, our only means for survival is to pass from one host on to the mucous membranes of an unsuspecting partner in love. The new surroundings hardly slow us down, and within a short time the cycle I just related is repeated. Yes, repeated worldwide at least a million times a year, making our eradication virtually impossible. We also love to make merry in joints and other tissue sites, but that is a story for another day.

"We express our appreciation to you for being able to share a small part of our story."

Sensing from the molecular nature of the various emissions from the assembled organisms that the assembly was getting a bit impatient, *Mycobacterium gordonae* called, "Colleagues, please hold your metabolic vapors for just a short time, for one more brief reading. The best is yet to come. I can assure you that your forbearance will be rewarded." This had the effect of clearing the vapors. The audience settled down, and *Mycobacterium gordonae* began its interpretation from the third plasmid.

Message of *Treponema pallidum*

"Fellow prokaryotes: I am *Treponema pallidum*, a somewhat sly, slick fellow as judged by humans, but I do express my appreciation for the opportunity to address you by proxy. I shall try to keep my account brief, although the disease called 'syphilis' is highly complex. In fact, syphilis is so varied in its manifestations that such terms as 'imitator' and 'great impostor' have been flung in our direction. We share with our colleague *Neisseria gonorrhoeae* the modus operandi of initially invading the genital tracts of humans, as well as the dubious distinction of being transmitted virtually at all times via intimate sexual relations . . . and not just coincidentally with the conjunction of Saturn and Mars, as certain primeval people once believed.

"Whence comes my name? You see I have inherited *nema*, or 'thread,' now coupled with the prefix *trepo*, which means 'to turn'—hence, the 'turning thread.' The name of our clan of most importance to humans is *pallidum*, which means 'pale,' though we would not be considered pale if early-20th-century humans had taken the trouble to find appropriate methods to mark us, instead of simply giving up when we declined to accept Dr. Gram's stain.

"The word 'syphilis' too is very interesting. Can you imagine that humans would actually name a disease after a poetic hero? Well, that is exactly what they did. The poem 'Syphilis sive Morbis Gallicus' ('Syphilis or the French Disease') was penned in the 16th century by Girolamo Fracastoro, also known as Hieronymous Fracastorius, a physician and poet who was a colleague and compatriot of Copernicus. This poem vividly describes 'our disease' and the plight of a shepherd boy who presumably was the first infected. The name 'syphilis' itself is likely derived from the Greek word *siphlos,* meaning 'maimed or crippled,' an apt description of the third and last stage of this disease. Too, this reference to syphilis as the 'French disease' is interesting from our perspective. It appears that humans do not treat themselves any better in throwing names at one another than they do us bacteria. In the case in point, it is our observation that humans become a bit testy or uncomfortable with any disease associated with their intimate relations or fornication. It is their practice either to keep such events in secret or to point a finger at someone else. The 'Spanish disease,' 'Columbus's disease,' the 'Portuguese disease'—in effect, '*someone else's* disease'—we've been called it all.

"In addition to this poem, Fracastoro published his theory that contagious disease, specifically what he called syphilis, was caused by 'rapidly multiplying minute bodies that are transferred from the infector to the infectee by direct contact.' Wow! If it had not been for the alchemist and mystic Paracelsus, who threw cold water on the germ theory of disease simply because he was unable to see us, we might have been discovered in much earlier times. Yes, Philippus Aereolus Theophrastus Bombast von Hohenheim, in arrogantly renaming himself 'Paracelsus,' boasted to all humankind that he was above and beyond [*para*] the famous first-century Roman physician, Celsus. Well, he certainly was beyond us. Even if he had not changed his name, or had had access to the best microscope of the day, I doubt if he would have seen us anyway, strictly because we are so thin and nimble.

"Key to the recognition that we have arrived on the scene is the sudden appearance of a shallow, punched-out, painless ulcer, called a chancre, on or near the private parts, appearing approximately three weeks following a coital episode. It has been estimated that one million of our slender, undulating, spirochetal cells per gram of tissue are required to produce a chancre. But take note: only two or three of our cells, finding their way to the soft, moist skin of the copulating organ, are required to produce an infection. No wonder that syphilis has been so contagious through the years. With other bacteria, an inoculum in excess of several hundred cells is required before the symptoms of infection ensue; we are far more efficient.

"I propose an exercise to provide you microbes with an interesting ribosomal stimulation. Assume that only two spirochetal cells are inoculated from one sexual partner to the other. Next assume that the doubling time for each cell is 24 hours, or once per day. How many days, then, will be required before a chancre appears, again assuming that a mass of one million spirochetes is needed? I shall give you a moment to generate some gas to clear your porins . . .

"Okay, the answer to the puzzle:

> After day 2, there are 4 in aggregate,
> By the end of day 3, there will be 8;
> After one week, 128 will replicate,
> By a fortnight into 16,000 will generate,
> Until day 20, 1,000,000 a chancre saturate.

"Indeed, this is the same well-known three-week time delay between exposure and the appearance of the chancre that is mentioned in all clinical treatises of this disease.

"How then do we go about doing our work? Perhaps you are familiar with tropism, a word from the Greek *tropos*, meaning 'an affinity for' or 'a tendency to turn toward.' Although every microbe may not have a distinct calling, so to speak, and may not be necessarily actively drawn to a particular tissue site or nidus to produce infection, most have preferred arenas of activity. You have just heard how *Mycobacterium tuberculosis* has a tropism for the respiratory tract and *Neisseria gonorrhoeae* for the genital tract. We *pallidum* treponemes have an obsession with the invasion of small blood vessels. No one has yet explained the phenomenon, but this trait, buried deep within our genes, results in the myriad disease syndromes that we 'great imitators' can produce.

"Our physiognomy may have something to do with it. We are as thin as a rail, and all coiled up like a corkscrew. An internal flagellum—we are, after all, spirochetes like *Leptospira*—provides us with an undulating motion that allows us to wriggle into tight spaces. Once introduced into human tissues, we make the outer layers of small blood vessels our first target. We then continue to bore through the vessel wall, unstoppable by the plasma cells that cartwheel to the fray, and finally after several days reach the inner (intimal) lining of the lumen. Ultimately the vessel becomes entirely obliterated, shutting off the supply of blood and leaving a narrow concentric zone of tissue death and fibrous scarring.

"Infected patients cannot be considered cured once the primary chancre heals. A diffuse macular or pustular skin rash marks our progress throughout the skin, prodromal to what humans have called the secondary stage of infection. What it's called is of little moment to us as we penetrate into virtually every organ, including the brain. Then follows the third and final stage of disease, the failure of the nervous and vascular systems, which leads to a whole host of protean psychosomatic signs and symptoms. Observing humans with personality changes, loss of judgment, seizures, slurring of speech, shooting or lightning pains, impotence, loss of hair, a staggering gait—any of these and many more will quickly hint that we spirochetes have been hard at work. The aorta may even blow up like a balloon, causing an explosion of blood as the bubble bursts.

"Where we came from and how this all started is enmeshed in mystery and folklore. Some modern-day humans prefer to believe the early writings of Aztruc (John Aztruc, *A Treatise of Venereal Disease*, 1754), who directly blames Christopher Columbus and his sailors for picking up our disease in the New World and bringing it back, singlehandedly causing the late-15th-century outbreak in Europe. Or they lean to the theory of Buret (Frederic Buret, *Syphilis in Ancient and Prehistoric Times*, 1891) that syphilis, under all sorts of other names, has been known in the Old World from very ancient times. Although we treponemes side with the latter, based on the information coded in our chromosomes, we cannot discount the possibility that a significant mutation in our ancestors' virulence occurred about the time of Columbus.

"It cannot be denied that evidence is strong that many famous and infamous humans, reaching back millennia before the modern age, had signs and symptoms indicating habitation by our treponemal ancestors. When one reads that a historical person 'suffered from the pox' or had recurring fevers, muscle and joint pains, particularly 'shooting pains' in the limbs; unsteadiness of gait; or a variety of personality traits such as dulling of the intellect, loss of insight, poor judgment, or overt cruelty bordering on insanity, the ravages of syphilis can be suspected. So too with any history of repeated miscarriages and stillbirths. Death in childhood, or descriptions of telltale signs in individuals such as saddle nose, narrow, notched incisors or mulberry molars, protruding mandible, or elevated ridges where the skull bones come together, all could indicate congenital and infantile syphilis.

"Note as one example the description of Christopher Columbus's last days (cited by Thomas Parran, *Shadow on the Land, Syphilis*, 1937). No matter where he acquired us, it certainly sounds as though he suffered from our devastating effects: 'He began to hear voices and to regard himself as an ambassador of God [now there is a personality quirk!]. After one last voyage to the New World, he had to be carried ashore. With his whole body dropsical from the chest downward, like that which is caused by injury to the valves of the heart, his limbs paralyzed, and his brain affected [all symptoms of late, fatal syphilis], he died.'

"And consider also consider the following account of King Henry VIII of England (as recorded by Dickson Wright, a London physician, in an unpublished manuscript acquired by Theodor Roseburg): Henry suffered from ulcerations of the legs above the knees, which did not respond to hundreds of remedies. He became inhumanly cruel in later life, executing an inordinate number of people he considered enemies, including two of his wives. Among his victims was Cardinal Thomas Wolsey, Henry's closest confidant (who, incidentally, admitted having the 'pox'), who was beheaded after being accused of infecting the king by whispering in his ear. Catherine of Aragon, his first wife, bore him six children, five of whom were stillborn or died shortly after birth. Her only surviving child, Mary—later Queen Mary I or 'Bloody Mary'—was described as being very thin, with moth-eaten hair, square head, protruding forehead, bad eyesight, and an erratic temper, and she died suddenly at age 42 (from a ruptured aortic aneurysm?), which all sounds very much like congenital syphilis.

"There are many more cases that can be cited, my good colleagues. However, with only circumstantial evidence at hand, and the distress that any such formal accusations might cause in human circles, it is best that we leave our disease history open for eternal speculation. My allotted time has expired, and I am most pleased to have shared a bit of our story with you."

Ignoring the waving of flagella, reflections of light off pigmented colonies, and an ouburst of metabolic odors, all vying for its attention, *Stenotrophomonas maltophilia* concluded, "To relate further accounts here would serve no particular purpose, and it is time to move on to our next session." With that, it descended from the podium, signing for Genie Transposon to place the Session 4 agenda on the podium kiosk.

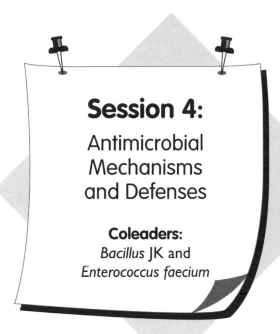

Session 4:
Antimicrobial Mechanisms and Defenses

Coleaders:
Bacillus JK and
Enterococcus faecium

Thermotoga maritima

Allow me to return briefly with some background information concerning the antibiotic onslaught being waged by humans. It was only through a fluke of circumstance, more correctly the lucky mishap of one human, that the phenomenon of antimicrobial action was recognized.

Over eons of the natural competition for habitats and nutrients, some fungi, Bacillus *species, and funguslike bacteria developed what humans call "antibiotics"—substances that kill or inhibit the growth of other microorganisms—to gain an advantage over their microbial competitors. In return, we prokaryotes have either simply avoided these organisms or developed mechanisms to neutralize their toxic effects. It is a sad irony that when the early human pioneers in the antibiotic movement began to manufacture their new drugs, they used many of these same species! No wonder that many of these medicines didn't have the desired effects—we had already developed resistance against these substances over millennia of cohabitation.*

Now for some history. There was once a superb scientist, Alexander Fleming by name (later knighted Sir Alexander). One day while working in his research laboratory, he forgot to cover a petri dish on which colonies of our colleague Staphylococcus *were growing. By luck, some airblown spores of a fuzzy mold called* Penicillium notatum *landed on the surface of the agar immediately adjacent to these unsuspecting staphylococci. Next day, Dr. Fleming noticed a clear, concentric zone around one of the* Penicillium *colonies, where growth of the* Staphylococcus *colonies was impaired. Impressively, he deduced that the mold was manufacturing a substance that appeared to be lethal to the* Staphylococcus.

As fortune had it, it was a "susceptible" strain of Staphylococcus *growing on that plate. Had that culture dish been occupied by strains that had become resistant by being for many years in close association with* Penicillium *in the wild, the phenomenon of antibiosis would not have been discovered! But some such discovery was inevitable, and it was luck, and Dr. Fleming's excellent deductive ability, that caused it to happen then.*

On the basis of Dr. Fleming's observations, people began to reason that if extracts from certain molds could be used to kill or inhibit bacteria, then such extracts might actually be used to cure human infections. Humans began to boast that we prokaryotes would finally be conquered, and a new era free of pestilence was about to begin! But elation soon gave way to reality and disillusion. Some infections were not affected by their magic potion, now called "penicillin." You see, in all their excitement, humans never stopped to ask us. We could have told them it would only be a matter of time before we found ways around the effects of their splendid new drugs. We've lived for ages in harmony with Penicillium *in nature— why would things be any different at the site of an infection?*

As time went on, more and more of the infection-producing bacteria exhibited resistance, forcing scientists to seek additional types

The Other End of the Microscope

of antibiotics. Human inventiveness in discovering new agents has been heralded as one of the marvels of the modern age. However, equally as ingenious have been the ways in which we prokaryotes have, in turn, evolved mechanisms of resistance against most of these new agents. For example, not long ago humans were boasting that tuberculosis was near extinction, yet, as noted in the previous session, they are faced now with inventive mycobacteria that are more resistant to their new drugs than ever before.

This is the focus of these presentations, and it is now time to begin. Enterococcus faecium *and* Bacillus *JK are ready to commence.*

> Man's discovery of antimicrobial agents must surely rank as one of his greatest triumphs. It is essential, however, that no one should underestimate the cunning of bacteria, who with remarkable ingenuity have developed methods of resisting the action of these substances so that some have already been rendered useless, whilst others are in the great danger of it.
>
> —Peter Baldry, *The Battle Against Bacteria*, 1965

Enterococcus faecium, its round spheres in short chains, and *Bacillus* JK, with its long boxcar forms—two microbes noted for acquiring clever defenses against antibacterial drugs—in turn ascended the podium. The excitement among the crowd now gave way to refocused attention on the speakers, in anticipation of this long-awaited session on the competition between human-invented antibiotics and the many means of self-protection devised by the bacteria.

Address of *Bacillus* JK

"My fellow prokaryotes! I am honored to be selected to present these opening remarks. It may well be that I am not the best qualified among you in terms of developing ingenious mechanisms of resistance; my clan's recent ascendancy to the top of the antibiotic resistance list is due more to pure circumstance than to any cleverness on our part. But I will do my best to speak for the 'newly emerging' antibiotic-resistant organisms—*Stomatococcus*, *Bacillus cereus*, *Leuconostoc*, *Rhodococcus*, *Stenotrophomonas*, *Moraxella catarrhalis*, and *Burkholderia cepacia*, among others—as well as for those who have long been recognized by humans.

"First, let me say that I prefer to be known simply as *Bacillus* JK, even though my official name in human circles is *Corynebacterium jeikeium*. Researchers originally isolated my clan as a curiosity: a newly recognized strain of *Corynebacterium* with some resistance to antibiotics. Being just one insignificant isolate among many, we were tagged with the simple code 'JK.' However, once interest grew in our unique resistance abilities, the taxonomists—those label-loving humans we all know so well—decreed that 'JK' was too crude and enlarged it into the comical tag *jeikeium* that now serves as our official species name.

"To follow up on what has been previously related, we '*jeikeiums*' are a group that for ages have shared our home in the soil with many of you and your closely related cousins. Inherently we are quite benign fellows, carrying out our brief existence with no intent to harm anyone. As with most soil dwellers, we have since the beginning of our times been faced with the need to deal with the various noxious and at times lethal substances either emanating from decaying matter or produced by our neighboring microorganisms. Such adaptation has been at the expense of millions and millions of our mates, but, fortunately for us, among each generation were always a sturdy few that had learned to alter their structure or shift their internal mechanisms in order to evade the lethal effects of these natural germicides. Silently, and virtually invisibly, we have either changed the codes in our chromosomes, or have assimilated DNA in plasmids and transposons from antibiotic-resistant colleagues, all strictly as a means of survival in a hostile environment: just one more form of natural selection.

"What we recognize as 'adaptation for survival' humans refer to as 'resistance,' and categorize us as opportunistic villains. But how can our adaptability be construed by humans as predatory?

"Consider *Moraxella catarrhalis*—formerly known as *Neisseria catarrhalis*—as an example. In its former days, this prokaryote happened on the human oropharynx as an ideal habitat. In time it became a commensal resident of the upper respiratory tract, causing no harm and passed off by human doctors as 'normal flora,' just part of the territory. However, recently *Moraxella catarrhalis* found itself fighting for survival against the onslaught of a battery of drugs that humans were using to combat the ill effects of our colleague *Streptococcus pyogenes* and other oropharyngeal bacterial residents. We perhaps can coin a new law of nature: 'The annihilation of one breed only stirs up anger in another.' What were the consequences? Certain strains of *Moraxella catarrhalis*, once relatively harmless, assumed a new coat of arms. These newly resistant strains took up residence in the lower reaches of the human respiratory tree, where they are now making their presence felt in the form of bronchial infections and pneumonia, particularly in folks who have chronic lung disease or who have diminished lung resistance from a lifetime use of cigarettes.

"As I mentioned earlier, we JK bacilli belong to the soil, but we also do quite well in the moist dirt and dust that accumulates in houses and public buildings, particularly where careful custodial care is lacking. In most settings we tend to lie low, minding our own concerns and having no effect on healthy humans. But now visualize what humans have created in the hospital setting. Obviously, this is a place for sick folks, debilitated people whose ability to deal with a new bacterial challenge may be weakened. So what happens? We may be suddenly disrupted from our dusty corner, whipped into the air in the draft as a nurse or doctor quickly passes by, and land on some unsuspecting patient. Even less excusable is the bus ride we enjoy when a health worker ferries us from one to another patient on his or her unwashed hands, or when we get transferred on some communal implement that they fail to clean.

Once in place, we take up residence on the skin or in the respiratory tract or gastrointestinal tract, wherever we can survive. We may gain entry into the deep tissues at the site of a puncture or surgical wound, ride into the urinary tract along with a deeply probing catheter, or be pushed into other organs with aspiration needles and probes. From these sites we may even gain access to the bloodstream. In many instances, our unfortunate victims have no defenses of their own against us. The drugs they are treated with have no effect on us, and they become seriously ill. But how can we be held responsible for this? We didn't choose to be there!

"Certain humans in the medical community may finally be getting the picture. Some are realizing that the use of an increasing array of antimicrobial agents only makes us more determined to find ways around the scourge. The practice of dumping literally shovels full of antibiotics into animal feed, for instance, is now being questioned, with people demanding whether the ring of the cash register from the sale of plump chickens and pigs is worth the cost of new antibiotic-resistant infections in animal caretakers and others who come in contact with contaminated animals or their products.

"So we, the JK bacilli, among several of our bacterial colleagues, are under fire. We are captured from the blood of debilitated hospitalized patients, usually those who have been overtreated with antibiotics. Most of these antibiotics do not affect us, but rather eliminate many of our microbial neighbors who might otherwise tend to keep us in balance. Once they are gone, we spread joyously unimpeded into our new surroundings. When an infection occurs, we are plucked out on culture plates, and then humans are dismayed when their tests disclose no antibiotic with which we have not already come to terms through previous exposure. We are labeled 'multiresistant'—definitely a pejorative term on the doctors' part, but perhaps one we can wear with a sense of accomplishment."

From the audience wafted a strong atmosphere of enthusiasm in support of *Bacillus* JK's conclusion. The bacillus bowed in

The Other End of the Microscope

acknowledgment, and resumed rhetorically, "Why are we, the JK bacilli, and so many of our colleagues able to withstand the attack of the infinite variety of antibiotics that are launched against us— that is, how have we become 'multiresistant'? That will be the focus of the stories you are about to hear.

"We bacteria have ingeniously evolved numerous defenses and, indeed, counterattacks against the antibiotic armamentarium that humans continue to invent. Yes, there have been many setbacks. But given time, we have been able to adjust. In fact, some of our colleagues continue to outdistance all human efforts, to the point that no effective antibiotic exists to treat certain bacterial diseases.

"This state of affairs has resulted in considerable angst among humans, but this doesn't bother us in the least. 'Bacterial drug resistance has become a worldwide problem!' blare the newspapers, magazines, and scientific journals. If humans developed a worldwide communication system by which they could keep closer track of our activities, better to predict and manage outbreaks in every nook and cranny, that might cause problems for us down the road; however, I am confident that we will continue to share DNA information with our colleagues, maintain our rapid reproductive prowess, and mutate at rates sufficient to counter all future human antibiotic assaults and ensure our species' survival.

"But enough of this philosophizing. I am honored and pleased now to introduce *Enterococcus faecium*, who will moderate the discussion of how many of the prokaryotes have developed a variety of mechanisms to 'cover our flagella,' to speak colloquially."

Address of *Enterococcus faecium*

"I am most appreciative to my good friend JK for providing this broad background against which we can paint a complete picture. Because of the intricacies of antibiotic and counterantibiotic actions and reactions, it is well to 'not lose ourselves in the trees'— I believe that is how humans put it. We must take precautions not to become enmeshed in irrelevant details. For my part, I will attempt to make as clear as possible how I and other members of our *Enterococcus* clan have been able to withstand the antibiotics thrown at us. In keeping with what JK has already shared with us, we do not claim any particular credit for developing these resistance mechanisms, although plasmid insertions from colleagues along the way have certainly enhanced our defenses. It is simply our nature to adapt.

"Certain of our abilities to withstand the action of antibiotics were undoubtedly inherited millions of generations back—borne in the chromosome, if you will. Humans realized early on that we would not just shrivel up under the attack of the original crude extract of penicillin, as was the case with most of the original strains of *Staphylococcus* and *Streptococcus*. In fact, just after researchers invented a way to measure the relative resistance of bacteria, a technique they call 'antibiotic susceptibility testing,' it was discovered that certain of the antibiotic-impregnated paper disks used in the test had no effect on our growth. We enterococci grew right up to the lip of the penicillin disk. Confused, scientists early on singled us out from other 'susceptible' streptococci, calling us the 'penicillin-resistant streptococci.' Only later did they realize that we really do not belong to the *Streptococcus* clan after all.

"Now, to minimize confusion, I suggest we refer to an outline of the various mechanisms we will be discussing. There are several ways in which antibiotics serve to disable or kill us prokaryotes. We can all take pride in the fact, however, that one or more of our bacterial colleagues have been able to evolve mechanisms to counteract nearly all of these, as follows." At a gesture from a short chain of *Enterococcus*'s cells, Genie Transposon leaped up to the podium kiosk and posted the list.

"Please do not be intimidated by this complex array of defenses," continued *Enterococcus faecium*, as those assembled read through the contents of the poster. "Each will be explained in some detail during the course of the following presentations. Many of you bacteria have already utilized one or more of these

Mechanisms of Antibiotic Action and Bacterial Resistance

- Penicillin-binding proteins—*and* their alteration

- Beta-lactams—*and* beta-lactamases

- Invasion through outer membranes and porins—*and* changing the locks

- Stalling ribosomal assembly—*and* modifying the ribosomal plant

- Stalling DNA gyrase—*and* spinning the wheel

- Disruption of metabolic pathways—*and* new detours

mechanisms and may be confused only by the human terminology. Realizing that most of you remain prey to human antibiotics, you may wish to arrange for conjugations with colleagues who can supply you with a DNA plasmid infusion to provide some protection.

"We shall follow the outline of topics in order—that is, as much as is possible in these microbial circles, where as you know every function in our cells is linked to every other. I have also asked *Bacillus* JK to return at the conclusion of this session to summarize the current status of our fight for survival against antibiotic actions, including a reflection on where the next challenges appear to be arising. Please feel free to raise questions and comments as we go along. The free exchange of information has always been an important key to our survival, as you well know from *Escherichia coli*'s discourse on plasmids."

Penicillin-binding proteins—and their alteration

"To begin with," *Enterococcus faecium* commenced, "the first topic, penicillin-binding proteins, requires a short review of our cell wall structure.

"A principal strategy of antibiotic attack is to prevent us from constructing our cell walls. Recall if you will the model presented by *Micrococcus luteus* of the gram-blue, or gram-positive, cell wall. As you have learned, we gram-blue bacteria have cell walls constructed from cross-linked sugar and protein moieties, the peptidoglycan (amino acid/sugar) or tessera layer. Humans have unraveled the details of this structure, and on this basis have discovered a tool that they hope will interfere with its formation: the beta-lactam antibiotics, the most notorious of which is penicillin. These antibiotics come equipped with a chemical lasso or snare, called the beta-lactam ring, as part of their molecule. This ring attaches to certain proteins, which humans call 'penicillin-binding proteins,' in the peptidoglycan grid of our walls, much the way a looped rope will hitch a boat to a post on the dock. Once looped into our proteins, the beta-lactam rings block our ability to assemble those proteins into their correct links in the peptidoglycan mesh. The structurally weakened cell wall begins to crumble. As *Micrococcus luteus* described so eloquently, osmotic pressure begins to build up inside the cell, and without its supporting structure, the cell ruptures and dies.

THE PENICILLIN COWBOY

"However, some of us gram-blue bacteria have devised an ingenious system to foil the beta-lactam lassos. I would like to ask our close cousin *Streptococcus pneumoniae* now to explain this device."

Streptococcus pneumoniae, caught a bit off guard, nevertheless was able to take advantage of its slick sugary capsular coat to slide hastily into position to make a response. "Good day to everyone," it began. "You need not be so formal. Humans call me 'Strep Pneumo'—if they can be so unceremonious, certainly you, my friends, are welcome to be. Indeed, we have had a long acquaintance with humans, probably for centuries, although only recently were we recognized as the cause of a unique type of 'lobar pneumonia' that generally remains confined to only one lobe of the lung.

"Now, to our topic. Picture again how penicillin, with its beta-lactam ring, lassos the penicillin-binding proteins in our cell wall, bringing construction to a halt. How did we learn to deal with those beta-lactam cowboys?

"One of the major defects in our defense against penicillin was that we built these penicillin-binding proteins in only one configuration—square. This made it easy for the ring of the beta-lactam lassos, which is also square, to slip right over them and bind firmly. One watershed day, however, some brilliant Strep Pneumo ancestor happened on a unique means of thwarting the lasso effect of penicillin. It simply made its penicillin-binding proteins round—or rectangular, or even hexagonal. The result? The square lasso would not fit!

"And so, through the slow process of sorting and resorting genes and plasmid DNA codons, we of the current generation have discovered how to make penicillin-binding proteins of all different shapes. The remaining proteins with a square configuration are still snared, but those with an odd-shaped construction are not. Thus, the action of beta-lactam antibiotics against us will never be complete. We have either a high level of resistance, if most of the binding proteins are so altered, or varying intermediate levels of susceptibility or resistance, directly proportional to the percentage of odd-shaped binding sites we display. Cell wall construction goes on, and even if not perfect, is of sufficient integrity to ensure our survival.

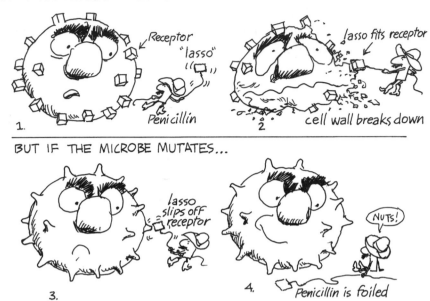

HOW PENICILLIN ATTACKS A MICROBE

Receptor "lasso"

Penicillin

1.

lasso fits receptor

cell wall breaks down

2

BUT IF THE MICROBE MUTATES...

lasso slips off receptor

3.

NUTS!

4. Penicillin is foiled

"Humans continue to scurry around trying to find alternative antibiotics to circumvent our clever manipulations, while patients are increasingly being put in jeopardy. And meanwhile, we have passed this plasmid-based survival information on to many of our colleagues, particularly to *Haemophilus influenzae* and *Neisseria gonorrhoeae*, spreading the good word, if you will."

An effusion of appreciation and encouragement wafted upward from the audience, in approval of *Streptococcus pneumoniae*'s ingenuity and generosity. *Enterococcus faecium*, having no flagella, rolled back and forth to gather the participants' attention.

Beta-lactams—and beta-lactamases

"Our thanks to *Streptococcus pneumoniae* for its contribution," it said. "As you know, colleagues, there is more than one way to deal with the beta-lactams. As explained, some of us can change the

The Other End of the Microscope

proteins that they target. And others of us attack the beta-lactams themselves.

"Of great interest, and of equal effectiveness, has been the ability of certain bacteria to produce substances that can break the beta-lactam ring, or cut the lasso, if you will. It may have been an ancient strain of *Staphylococcus aureus* that first produced these magic scissors, either accidentally or in response to antibiotic pressures in its soil habitat. In any case, today many bacterial strains possess this beta-lactamase—the 'ase' suffix meaning that it acts like an enzyme, snapping the beta-lactam rings not just of penicillin but also of a number of other beta-lactam antibiotics.

"Picture it, friends. We are minding our own business—albeit causing an infection in the process—and suddenly find ourselves facing a swarm of penicillin molecules, charging our way with their lassos at the ready. At first we may be caught by surprise, but within moments our ribosome factories are stimulated to manufacture huge amounts of these beta-lactamase scissors. They pass readily outward, through our inner cell membrane and the loose peptidoglycan wall, to surround our cells with a thin cloud. Although some of the initial penicillin molecules may permeate into our peptidoglycan layer and even snag some of our cell wall binding proteins, they often are too few to overcome the strong defense of our beta-lactamases. Lasso after lasso is snipped apart, and, unable to anchor to our cell wall building blocks, the penicillins drift harmlessly away. We quickly repair any cell wall damage done by the initial onslaught and carry on unabated with our infectious process."

"Now, for another view of this same conflict, let me recognize our friend *Haemophilus influenzae.*" *Enterococcus faecium* rolled briefly forward on the podium, signaling that *Haemophilus influenzae* was welcome to begin.

"Fellow prokaryotes," the new speaker commenced. "For those of you who do not know me, the name humans have given me is *Haemophilus influenzae, Haemophilus* meaning 'blood lovers,' or 'vampires' that need blood to survive. You probably have heard of me as a renegade that causes meningeal infections (among other maladies) in humans, particularly in children. Part of this effect is owed to our ability to produce beta-lactamases and resist the efforts of humans to drug us out of existence. But my reason for participating now is to relate the method that we gram-red bacteria use to combat the effects of the beta-lactam antibiotics.

"Please remember that we, the gram-negative or gram-red bacteria, do not possess the thick peptidoglycan layer that houses our good friends on the gram-blue side of the spectrum. As explained in the second session by *Enterobacter aerogenes*, we have instead a thin outer membrane, only a few peptidoglycan units thick, like a plastic wrapper. Our contact with the outside is through special proteins called porins, as *Enterobacter aerogenes* described so clearly at that time. These porins act as miniature channels through the outer membrane. It may be difficult for some to visualize, but each tiny gram-red bacterial cell may have many thousands of porins! Penicillin molecules actually find it more difficult to reach inside our cells to damage us, thanks to the tight security checkpoint of our porins.

"Now, you must recall, between this outer membrane and the thin, phospholipid-rich cytoplasmic membrane that encases our inner cytoplasm and organelles is a region called the 'periplasmic space.' Any substance that has managed to pass through the outer membrane channels of the porin proteins must still pass through this periplasmic space before it can reach the inner cytoplasmic membrane.

"The inner cytoplasmic membrane has a unique property of being able to accumulate and release a variety of beta-lactamases, much as defenders with drawn arrows might line the inner castle wall. As each penicillin molecule, with lassos at the ready, passes inward through the outer membrane and enters the periplasmic space—ping, ping, ping! One by one the lactam rings are snapped. With no place to anchor, the fragmented penicillin pieces degrade into nothingness. This is the method we have devised to save ourselves from the beta-lactam antibiotics, so that we can continue our existence."

The Other End of the Microscope

"Thank you," replied *Enterococcus faecium*. "We appreciate greatly your eloquent explanation of gram-red structure and defense. This illustrates even more the high level of sophistication at which we bacteria can operate.

"Before leaving our discussion of the beta-lactamases, however, let me relate one last development that is of some concern. Humans indeed are clever. Soon on the heels of our discovery of how to manufacture beta-lactamases, they devised a counter-move to inactivate that system of defense.

"In short, they have constructed substances called clavulanic acid and sulbactam (i.e., beta-lactamase inhibitors), which appear to us to be identical to the beta-lactam antibiotics (penicillins, for example). These compounds are now combined with various beta-lactams into new drugs of great potency. This is bad news, because we cannot distinguish between the true antibiotic molecule and these look-alike compounds. When we send out our beta-lactamases, they attack the impostors and not the true beta-lactams. This leaves our penicillin-binding proteins vulnerable to attack by the remaining beta-lactam lassos, just as before.

"This has been somewhat of a setback for us, as I'm sure you will agree. However, we prokaryotes are nothing if not ingenious. One or another of us will surely come up with a new mechanism to counteract this dangerous development."

Enterococcus faecium then noticed *Chryseobacterium meningosepticum*, in its bright yellow coat, agitating for attention. The moderator nodded in assent.

Spontaneous Effusion of Chryseobacterium meningosepticum: *Holes in the Dyke*

"Fellow bacteria. I am *Chryseobacterium meningosepticum*. You may better recall me as *Flavobacterium meningosepticum*, before my name was unceremoniously changed by some human afterthought. We have just happened on some information that should be known by many of you gram-red bacteria. Polymyxin B is a polypeptide antibiotic that has a corkscrew-like action, boring holes through your inner plasma membranes. You can imagine what happens then! We want to let you know of this mode of antibiotic action, because humans may elect to use against you a combination of polymyxin and tetracycline. What does this gain them? Those of you who feel protected from tetracycline because you have developed an outer membrane barrier must realize that, when polymyxin is also thrown in, large holes can be drilled in your hide, potentially allowing passage of several antibiotics that you once could block, including tetracycline.

"We flavobacters have developed resistance to this cork borer process by tightening up our lipoprotein matrix. Unfortunately, our ability to resist this action is buried in our chromosomes and therefore we cannot pass the trait on to any of you. You may wish to instruct your genetic engineers to take inventory to uncover any stray DNA fragments circulating around in some untapped plasmids that may be of help to you."

"Thank you for this warning," replied *Enterococcus faecium*. "It is indeed timely and appropriate. We at this assembly should all be aware that, as always, we prokaryotes can never relax our vigilance. The ingenuity of humans is almost as infinite as ours. They have discovered, as *Chryseobacterium* says, that we bacteria can be better conquered if they bring into play more than one mode of antibiotic assault. 'You hit 'em high and I'll hit 'em low,' as a football player might say. Humans call this 'combined' or 'synergistic' antibiotic therapy. Although we may withstand the cell wall–damaging effects of most of the beta-lactam antibiotics by shifting our penicillin-binding proteins, as described before, this protection is not complete. Many of the higher-powered cell wall–active agents have been able to sufficiently weaken our outer protection, or eat holes in our coat, so to speak, that a sufficient number of the larger antibiotic molecules can pass literally uninhibited. This is a hazard that needs serious consideration in our evolutionary mutations; perhaps we should direct more attention to altering our ribosomal binding sites, as will be discussed later.

"Now, are there any other contributions before we turn to the next topic on our list?"

"Your pardon!" came a reply. *Staphylococcus epidermidis* was signaling for attention from its position on a bit of plastic that had somehow blown into the pond. "Friend moderator, while we're on the subject of new antibiotics with regard to our cell walls and cell membranes, should we not discuss the challenge of vancomycin? That story, at least, contains some heartening news."

"An excellent suggestion," *Enterococcus faecium* replied. "Please speak, *Staphylococcus*, as we know you and your cousins have been dealing with that hazard for some time."

Spontaneous Effusion of Staphylococcus epidermidis: Reconfiguring the Alanine Bridges

"I thank you." The bacterium slithered a little way down the slime layer it had formed on the plastic, to reach a better position to be heard. "Fellow prokaryotes, most of us here share in a long communal history of contriving counterattacks against antibiotics. I wish to share one recent victory. Vancomycin has for some time been a drug difficult for us to counteract. I am happy to report, primarily on behalf of my near cousin, *Staphylococcus aureus*, that it is beginning to mount some resistance to the action of this long-time nemesis. Thanks to key plasmid genes our cousin has shared with us, we too are having some success.

"Vancomycin is a large-molecular-weight protein/sugar complex called a glycopeptide. Its properties are rather different from those of the beta-lactam antibiotics, which means that our beta-lactamase countermeasures have no effect on it. Vancomycin also interrupts gram-positive or gram-blue cell wall synthesis, but at an earlier step, during the construction of the cross bridges—the alanine-alanine struts—that stabilize our peptidoglycan strands. Vancomycin binds with these alanine-alanine bridges, preventing cross linkages. Locking up these bridges is similar to preventing a carpenter from hammering in the cross braces of a frame wall. Thus, the cell wall is extremely weakened at the early stages of construction.

"This is where a colleague in the *Staphylococcus aureus* clan has recently come to the rescue. Presumably through the acquisition of a stray plasmid carrying a novel gene, some few *Staphyloccus aureus* cells learned to manufacture an enzyme that substitutes a lactate molecule for one of the alanine moieties. The resulting alanine-lactate cross struts, used instead of the conventional alanine-alanine link, do not affect the strength of our cell wall. What's most important to us, however, is the fact that vancomycin doesn't bind to these new alanine-lactate targets. We are resistant again! After several generations of replication, a substantial number of these alanine-lactate–producing, vancomycin-resistant staphylococci have emerged, again generating much anxiety among humans."

"Thank you, *Staphylococcus epidermidis*," replied *Enterococcus faecium*. "Indeed, what you have related about vancomycin is very true. We share both your concern and your elation. Our *Enterococcus* clan too have developed vancomycin resistance, to the point where humans now recognize a group of us as the VREs, short for 'vancomycin-resistant enterococci.'

"Our mechanisms of resistance are similar to yours, but a bit more detailed. Various members of my clan have acquired three new genes that seem to regulate our ability to resist the action of vancomycin. Those of us who are gifted with the first of these genes, *vanA*, become very resistant to the action of vancomycin, probably by substituting all of our cross bridges with alanine-lactate struts, which virtually stops vancomycin binding. It has been speculated that someone in our ancestry acquired a transposon containing this gene coding for this amino acid substitution, which we have been able to preserve through millions of replications.

"A second gene, called *vanB*, codes for a different effect, namely, only partial inhibition of vancomycin binding. In these bacterial strains, many alanine-alanine bridges remain, so that sufficient vancomycin binds to partially affect the rigidity of the cell wall. Even in this wobbly state, a sufficient number of cells survive to maintain infection. But when high levels of vancomycin are administered, our *vanB* enterococci have no defense.

"The third gene, *vanC*, is of interest in that it is a chromosomal mutation limited to only two of our brothers, *Enterococcus gallinarum* and *Enterococcus casseliflavus*. These two species are a bit odd for our clan, in that they produce a yellow pigment and, above all things, are motile. Who ever heard of a motile *Enterococcus*? Must have picked up a flagella-controlling gene from somewhere. The *vanC* gene is actually a mutation in the chromosome of these two select species—'in their bones,' so to speak, rather than simply borrowed on a transposon from someone else. Since there are no *vanC* plasmids swimming about, transfer to other species is not possible. *vanC*-carrying strains display only a low level of resistance against vancomycin, indicating that most of the amino acid precursor cross struts are alanine-alanine."

Enterococcus faecium then turned to the list of topics posted by Genie Transposon on the kiosk by the podium. "It is now high time we move on in our agenda for this session. We have discussed a variety of antibiotics with action against our outer membranes and cell walls. There are, however, a vast range of both natural and human-invented antibiotics which act against our *interior* mechanisms. The prime defense against these compounds is, of course, simply to prevent them from passing through our outer guards to reach and damage our insides. We prokaryotes have fortunately developed a number of mechanisms to accomplish this. Our colleague *Shigella sonnei* probably has the most experience in these mechanisms, and thus, let me now introduce its presentation."

Invasion through outer membranes and porins—and changing the locks

Shigella sonnei extended all of its flagella as a sign of acknowledgment. "I thank you, *Enterococcus faecium*, for this opportunity.

"Indeed, one of our primary antibiotic defenses is basic in concept: simply block the passage of damaging antibiotics by cinching down or altering the configuration of the porins in our outer cell membranes. As our colleague *Haemophilus influenzae* mentioned earlier, each bacterial cell has many thousands of porins, or channels through our outer membrane. There are so many different types of porins, in fact, that humans have even named those that recur with some frequency—OmpC (outer membrane protein C) or OmpF, for example. Perhaps in its original or natural state the porin channel configuration was such that certain antibiotics were able to pass through, as was the case with tetracycline-sensitive gram-red bacilli. But the passage of tetracycline, or any other substance, through the membrane porins is not just a matter of falling into a hole, as one might fall into an open mine shaft.

"Molecules that are large or of peculiar shape obviously have a more difficult time passing. For example, the tetracyclines are rather oversized, cumbersome molecules that are relatively easy to block. Our *Shigella* engineers have studied in some detail the structure of the tetracyclines to determine the most efficacious means to mount a defense system. As the name tetracycline indicates, this molecule includes four benzene-type rings. Each ring has six carbon atoms, with side chains attached at various positions, providing for differences in the various analogues. These chains are placed in tandem, which, when coupled with the side chains, results in a molecule that is somewhat unwieldy and strung out. This makes it relatively easy for us to keep such cumbersome molecules from ever penetrating our outer membrane.

"Also, the electric or ionic charge of the bacterium and of the antibiotic can have a distinct effect. Since the phospholipids in the bacterial cell and cytoplasmic membranes are negatively charged, any molecule that also has a negative charge will meet resistance. In contrast, passage is facilitated for small antibiotic molecules that may have a neutral charge ('zwitterions' or dipolar ions), which leaves us and our bacterial friends vulnerable. Further, once in the periplasmic space, most antibiotics are 'pulled' across the inner cytoplasmic membrane because they hitch up with an anion ferry—a transporter called the 'proton motive force.'

"Now you see what is coming. How did the first bacterium—I believe it was *Escherichia coli*—become resistant to tetracycline? Well, it invented different ways to shut the molecules out. For example, through mutation, a microbe can eliminate specific porin configurations that more easily admit an antibiotic. Or the porin structure may be narrowed, or altered physically or electrically, to block entrance for the antibiotic.

"Finally, another major alternative method of resistance we have invented is to actually pump tetracycline molecules back out of the cell, even after they have passed in through the outer membrane porins. This 'efflux pumping' phenomenon is really a masterful evolution. It would be difficult for any human, observing the thousands of tiny 2-micrometer specks of bacteria through a microscope, to conceive that each one possesses such a tiny 'sump pump'! This is another example of how humans should never take us prokaryotes for granted. We may be little specks but we have the most intricate machinery, both for the manufacture of a wide array of products and in the mechanisms by which we can defend ourselves."

Stalling ribosomal assembly—and modifying the ribosomal plant

"We thank you, *Shigella sonnei*, for these encouraging bits of information," said *Enterococcus faecium*. "Your discussion of tetracycline and defenses against it, in fact, brings us to the next topic in our agenda.

"The aminoglycosides, including gentamicin, tobramycin, and kanamycin, for example, have as their main target the 30S subunit of the ribosome, where they bind and shut down our protein synthesis. Once our ribosomal protein factories are damaged or destroyed, we suffer a great internal famine; many of us die, and others are weakened to the point of helplessness against the various immune mechanisms of our human hosts.

"But not all is lost. We have developed clever mechanisms that help us resist the antibiotic actions directed against our ribosomes, to the point that aminoglycoside-resistant enterococcal descendants have now emerged as some of the more troublesome of the bacterial species—troublesome for humans, that is. Achieving this position has not been without work and much frustration. However, as we are not alone in this accomplishment, I now introduce our colleague *Serratia marcescens*, whose story of acquiring aminoglycoside resistance is not only instructive but entertaining as well. "

A few minutes were needed, in fact, to let some of the collected metabolic vapors diffuse, as most members of the assembly were in the steepest pitch of logarithmic excitement upon hearing this variety of successes and defeats in the struggle against human aggression. During this interlude, *Serratia marcescens*, with flagella flying, assumed the presenters' podium, quite energized by what it had heard and in anticipation of what it was about to relate. When the curve of metabolic activity began to flatten, *Serratia marcescens* began its discourse.

Spontaneous Effusion of Serratia marcescens: Polenta and the Weeping Madonna

"At this juncture of our assembly, please allow me to digress for a few moments to share with you a few interesting details of our story: we, the *Serratia*. Such a detour may be welcome at this juncture, as I realize that pure scientific discourse can become quite fatiguing. So let us lighten up a bit as we enter a brief period of story time.

"First of all, we *Serratia* have considerable interest in the human taxonomy issue coming up in the next session of this assembly. Some of you may believe that our name *Serratia* relates to onc of our unique properties that, when we grow on agar surfaces, just as we are on this pond, the margins of our colonies tend to be a bit irregular, or 'serrated.' Not so; here is the real story.

"Living in the mid-18th century was an Italian, Seravino Serrati, who considered himself a 'physicist' but in reality was more of an entrepreneur, amateur seafarer, and man about town. Signor Serrati, to our understanding, was not always the most honorable of characters. Nonetheless, he was involved in the invention and early development of a prototype steamship engine, for which, at the time, he received little recognition.

"Some years after that, one Bartolomeo Bizio, a pharmacist at the University of Padua, was asked to join a delegation of university professors investigating the claim by one of the local peasants that a batch of polenta, a popular cornmeal mush or porridge, had spontaneously developed a bloody discoloration. Was it a miracle? No. After due observation, Bizio concluded that this phenomenon was caused by the action of a microorganism, an observation that was not held in much credit by his fellow delegates at that early time in history. However, being single-minded and somewhat inventive, Bizio placed polenta mush into a warm, moisturized chamber exposed to the air. After several hours, a few wine-colored spots were observed in the mush, and within 24 hours they covered the entire surface.

"Bizio formally published these observations in a scientific gazette, concluding that the discoloration was 'due to the seeds of some microscopic fungus present in the air.' In this report, he named the 'microscopic fungus' *Serratia marcescens*, in recognition of the contribution to humanity made by his countryman and friend, Seravino Serrati, whose contributions to Italian science he felt had for too long been slighted. (Many of you, my good prokaryotes, have complained about having the name of a human attached to your name. But in most instances, your human namesake has at least been a scientist, often a microbiologist, who had made a legitimate contribution to microbiology. Not we *Serratia*! No, we are named after a sea captain, perhaps less than respectable, who never saw a bacterium in his life!) The second name, *marcescens*, taken from a Latin stem meaning 'to fade' or 'to decay,' was also taken from Bizio's original observation that the pigment faded in time after the polenta mush was exposed to light.

"As an interesting sidebar, Bizio was later asked to investigate the reports from a local church that the 'blood of Christ' had entered the Host and that 'bloody tears' were seen exuding from the eye of a marble Madonna gracing one of the sanctuaries. After the pigmented polenta experience, and particularly based on Bizio's recent publication, the investigators concluded that the answer to these phenomena was more than likely scientific rather than theological. Nevertheless, the word of these 'miracles' spread quickly throughout the land, and any scientific explanations were soon trampled under the feet of thousands of pilgrims. We the *Serratia* have no opinion in this matter, except to say that we do like to grow in polenta and we produce a water-soluble pigment that can discolor anything with which it comes in contact. We remain of the opinion that some practical joker among our ancestors probably occupied the Host bread and the eye socket of the statue just mentioned. Nonetheless, the 'miracles of Padua' remain chronicled to this day.

"Well, to continue the taxonomic silliness, late-19th-century microbiologists in other parts of Europe independently named us *Bacillus prodigiosum*. The name *Bacillus* recognized our long, slender form, and *prodigiosum*, from the Latin *prodigiosus*, meaning 'portent' or 'omen,' refers to the miracles in which we were allegedly involved. We might have preferred a name such as *Bacillus rubrum*, in recognition of the blood-red pigment so characteristic of our persona. 'The red bacillus' has a certain ring to it, much like 'The Red Baron,' don't you think? Or we would perhaps would have dubbed ourselves *Semolina rubra*, the former name in reference to bran or wheat flour, on which we were first discovered, and, of course, *rubra* meaning 'red.' Or, one of our cousins may have suggested a name such as *Farinacola rubiate*, that is, 'living in association with farina [another name for meal or flour], with a ruby-red glow.' But we were not asked, and thus modern humans know us by the name of *Serratia*, which has no connection to our appearance, functions, or anything microbiological.

"In reference to our ability to produce red pigment, we would like to relate yet one additional example of human folly. In the mid-20th century, as late as the 1960s, humans were attempting to get a better grasp, for epidemiological and infection control purposes, on where prokaryotes exist in their environment. They specifically wanted to know how we grow and spread, particularly in the hospital setting. Some bright soul conceived the idea that our beautiful red colonies might be a useful visible marker for tracking purposes.

"First we were trapped in a vapor mist. We really did not mind this treatment; in fact, we rather liked it since we thrive when moisturized, as in polenta as I have said. Next, we were 'misted' to the four winds, down hallways and into obscure rooms. The next day a follow-up observer swabbed material from floors, walls, and various inanimate objects and plastered us on the surface of an agar dish to learn about our travel history. Sure enough, the following day they quickly spotted our red coats, and entered *prodigious* notes"—the audience responded to this pun with a variety of amused or disgusted exhalations—"notes, I say, of their observations into a ledger. All they evidently learned from this burst of activity was that doors that do not reach the floor will leak air, allowing wide distribution of commensal and environmentally adapted microbes. With all their advanced acumen, one would think that humans would have figured this out before! The rationale of this exercise in futility remains a mystery to us to this very day.

"But to continue the story. We were used in this capacity because we were so nonthreatening; indeed, in our pigmented form, we were rather benign fellows, limiting our self-expression to producing beautiful red coats. But after the humans had spread us from one corner of the establishment to the other, we were abandoned. They even failed to spread a disinfectant or make any other attempts to mop us up, but left us to our own devices. Perhaps they had not reckoned at that time that we flourish very well at room temperature. We took up housekeeping in the crevices and crannies of these contaminated rooms with great delight. In this process we also came in contact with other 'dirt-loving' colleagues, some of which possessed plasmid traits that were distinctly opposite to our nature. We conjugated with these fellows, assimilating many of their antibiotic resistance genes and learning a new, aggressive persona. We flung off our red ornamentation and, with our new antibiotic-resistant coats, raised a proud new cry—'viva *Serratia*!'

"The antibiotic era was in full swing by then, and our ancestors and colleagues all over the globe were under siege. We the *Serratia* were among the first to thwart these antimicrobial efforts. With our newly acquired antibiotic resistance, we suddenly appeared on catheters, in open wounds, and anywhere else we could find a foothold. Most of the antibiotics of that era could not touch us. Once we gained access to the bloodstream, it was time to say goodbye. We soon became Public Enemy Number 1, and every effort was made to eradicate us from the environment. But without our red coats, we were much more difficult to spot and could hide out in various unlikely places. We were also difficult to find or to differentiate from look-alike colleagues in the laboratory culture plates.

"The blame that humans have put on us for being such unruly characters is both amazing and disconcerting as we look back on this period of our history. Given the opportunity to make our case, our stance would be that humans were victims of their own stupidity. The fact that we shed our red coats and acquired many ways to fend off their antibiotic attacks had nothing to do with a vindictive spirit. As lowly creatures of nature, we have neither 'brains' nor 'consciousness' with which to plan aggressive strategies, nor to judge the consequences. We can't even be faulted for conjugating with colleagues of ill repute; who put us in that position in the first place? For the sake of argument, what if a large number of domestic dogs were let loose in a wild forest near an imaginary town? When the town's chickens started to disappear, the villagers would probably blame the losses on a pack of wolves, or witches or something. Only in retrospect might they unscramble that their abandoned pets possessed feral tendencies that became expressed, for the sake of survival, while living in the wild.

"Now that our story is told, let's go back to the main theme. *Enterococcus faecium* has already introduced the aminoglycosides and their main target of action on the ribosomes. I do not know that we, the nonpigmented *Serratia*, can lay any particular claim to pioneering resistance to this group of drugs, but genta-micin and kanamycin were being used in great quantities during the time of that red dust caper. We certainly were in the middle of the fray and subject to extreme 'search and kill' operations. It is most likely that we acquired aminoglycoside resistance genes from one of our dirt-dwelling cohorts. Thus, it is best to consider merely that we were among the first to use conjugation as a front-line armament.

"We have two options to ward off this aminoglycoside threat, both of which we exercise. Either we can try to destroy or some-how disable the aminoglycoside molecule, or we can attempt to change our ribosome so that the drug molecule is unable to rec-ognize or attach to the binding site.

"First, we produce modifying enzymes that serve to inactivate aminoglycoside molecules, much as our lactamase-producing gram-blue colleagues disable the penicillins. But the rings, or the lassos, in the aminoglycoside molecule are too tight to cleave in the manner of the beta-lactamase. Rather, we produce phosphor-ylases, acetylases, or adenylases that attach small molecules to vulnerable sections of the main aminoglycoside molecule, taming its action much as a tranquilizing dart brings down a bellicose bear. Thus, the effectiveness of streptomycin and kanamycin has been so diminished that humans hardly use them at all. Unfortu-nately, gentamicin and tobramycin manage to ignore these inacti-vating enzymes and go about their destructive business much as before. Thus, we must look to ribosomal reconfiguration as a backup ploy.

"However, alteration of the ribosomal structure is a bit more complicated. As you all probably know, a particular configuration of the protein structure on the surface of the ribosome, known as a binding site, is the target for a complementary tag on the antibi-otic molecule. When a match is found, antibiotic binding occurs. For purposes of illustration, visualize a string of beads made up of a sequence of different colors—say, red, yellow, blue, and green. An identical sequence of complementary reactive groups on the aminoglycoside molecule—red, yellow, blue, and green—is required for binding. What then if, through chromosomal mutation or from the action of an acquired plasmid gene, we *change the sequence*—yellow, blue, red, and green? What is the aminoglycoside monster going to do? Not finding a match, it bumbles on its way, leaving us undisturbed. Now we are a resistant strain. How clever of us!"

"Thank you indeed, *Serratia marcescens*, for a most entertaining and enlightening presentation," concluded *Enterococcus faecium*. "It is encouraging that an increasing number of my enterococcal cousins have also acquired these clever ribosome-altering modes of protection. In fact, certain members of our clan have become so notorious they are called the 'highly resistant strains,' as I will explain.

"Just to briefly broaden the story: we enterococci for some time were smug in our ability to withstand the aminoglycosides by preventing their passage through the tangled peptidoglycan matrix of our cell walls—until humans invented combination therapy. This strategy takes advantage of the synergistic effect that penicillin has when it is added to an aminoglycoside regimen. The penicillin opens up our cell wall, allowing free access of the aminoglycosides to our ribosomal factories. However, as described by *Serratia*, by altering our ribosomal binding sites we have been able to withstand the effects of many aminoglycosides, no matter how plentifully they pass through our penicillin-weakened cell walls. Thus, many of us have become synergy resistant, or, as mentioned, 'high-level aminoglycoside resistant.'

"Now our time for this session is growing short. Two more methods of defense against antibiotics must be presented before we conclude. Next on the list is the resistance mechanism involving alterations in the DNA gyrase. I see that *Escherichia coli* is still here, waiting to moderate the next session on taxonomy. Good day, colleague! Please share with us how you and other members of your clan have been able to withstand the recent attacks of the quinolone antibiotics, which now present a serious threat to our well-being."

Stalling DNA gyrase—and spinning the wheel

"Greetings to all you resistance fighters!" *Escherichia coli* replied. "I am pleased to be asked to contribute. Indeed, the quinolone group of antibiotics do present a special challenge for us. A quinolone drug, you see, possesses only two carbon rings, one with a nitrogen substitution, which, even with some side chains, makes it a relatively small molecule. It also is not particularly hydrophilic, and therefore it easily transmigrates our outer membrane porin channels. After humans added fluorine-containing moieties to the side chains, the quinolones became even more effective against us—for example, our gram-blue colleagues, such as *Bacillus anthracis*, are inhibited by the fluorinated quinolone ciprofloxacin. At this time, fewer than 5% of our strains have been able to resist these drugs. This is largely due to our difficulties in either inactivating the drug molecules or altering their target. You see, the quinolones have been designed to affect the very core of our survival: our ability to reproduce.

"As you know, in order to replicate, we need to make a duplicate copy of our chromosome to inhabit the new cell. Normally we do this very efficiently: the replication or doubling time of bacteria can be as little as 20 minutes. (Compare this with the doubling time for humans—20 years or more!) We can go through up to 72 doublings, or 2^{72}, within a human day, achieving an enormous number of cells. Humans know this, and so they have invented antibiotics like the quinolones that block the activity of the enzymes we use in manufacturing our DNA. This literally stops us in our tracks.

"Our DNA manufacturing process is unequivocally amazing, more efficient than any production line established by humans. The master design lies in the DNA of our chromosomes. The ultimate origin of the bacterial chromosome and its structure, of course, is long lost to all but philosophers, but ours is little different from the chromosomes of all other life. Now, forgive me if I repeat information you know already; it will help to set the stage for understanding this entire process.

"Our DNA, or deoxyribonucleic acid, is a long strand composed of repeating molecules of the five-carbon sugar ribose, from which an oxygen has been removed ('deoxy') and which are linked in turn to one of four nitrogenous bases (adenine, guanine, cytosine, thymine) and to a phosphate group. This strand of molecules, or nucleotides, is referred to as a polynucleotide. Two polynucleotide strands, in opposite (head-to-tail) orientation, are weakly bound together by hydrogen bonds between specific complementary nucleotides, with adenine (A) always linking with thymine (T) and guanine (G) always linking with cytosine (C). This double-stranded unit is not stretched out like a ladder, but rather is twisted into a spiral or helix, much like a spiral staircase in a tower, a much more compact and yet sturdy structure. A series of enzymes called polymerases are involved not only in assembling the DNA polynucleotide strands unit by unit, but also in relieving the twisting tension, called negative supercoiling, which naturally builds up in the DNA helix as the polymerase progresses along the DNA template (gyrase).

"Gyrases have two subunits. Subunit A is responsible for breaking the sugar-phosphate backbone of one of the two strands of DNA, allowing the free end to swivel a full 360 degrees, reducing that supercoiling tension. Subunit B then reseals this break, bringing the DNA strand back together in its functional, double-helical structure. But the quinolones attach to receptors on the A subunit, which prevents the DNA from splitting. Tension thus builds up in the DNA, due to the inability to remove supercoils, and the progress of the DNA polymerase soon stops.

"But there is some hope. A few of our colleagues have encountered a plasmid containing a *gyrA* gene that codes for key amino acid substitutions in the gyrase A subunit. Once such a substitution is made, the quinolone molecule cannot attach, which neutralizes its effect and thus allows DNA synthesis to proceed unimpeded. This resistance mechanism has worked well against nalidixic acid, one of the original quinolones, but has been less effective against newer analogues that have different binding properties. Thus, we must be patient. As with the development of other resistance mechanisms, I am confident that it is only a matter of time until, under the increased human use of the quinolones, we will happen on a chance mutation or encounter a plasmid with the necessary gene."

"Thank you, *Escherichia coli*, for this guardedly optimistic report," said *Enterococcus faecium*. "Now, to address the final topic on the list of antibiotic resistance mechanisms—the use of alternative pathways of metabolism—*Bacillus* JK has asked me to relate briefly the method my clan has developed to resist the action of the sulfonamides."

Disruption of metabolic pathways—and new detours

"As most of you recall, the sulfonamides were among the first of the antibiotics with which we had to contend. Early on we were only concerned with resisting sulfonamide itself; however, later on humans added trimethoprim to augment its effects. But I will deal with that development later.

"Our resistance to sulfonamide is one of those jokes we have pulled on humans, illustrating that even we prokaryotes embody an element of the trickster. For a full understanding of how we handle the action of the sulfonamides, I must first describe a chain of events. Please give careful attention.

"The sulfonamide molecule is a small one, a single benzene ring with a free amino group attached to a carbon atom at one side and a sulfonyl radical attached to the carbon on the opposite end, a position termed '*para*.' (One might simply consider *para* as meaning 'opposite,' but in chemical language it means 'in a position separated by two carbon atoms'—which in this case, because there are only six carbon atoms in the benzene ring, turns out to be 'opposite.') The sulfonamide molecule is similar to *para*-aminobenzoic acid, a compound that is needed in the bacterial metabolic pathway to synthesize folic acid. Folic acid, in its turn, is essential for the synthesis of the nucleotide purines that compose DNA and RNA. Without folic acid, new DNA and RNA cannot be synthesized, bringing our manufacture of protein to a halt.

"Sulfonamide works undercover, by inserting itself in the nucleotide synthetic pathways where *para*-aminobenzoic acid would normally be at work, shutting down the synthesis of folic acid. Luckily for us, this halt in synthesis is only temporary, a transient condition that humans call a 'bacteriostatic' effect. Once the concentration of sulfonamide is reduced, folic acid once again is produced, so that DNA and mRNA synthesis and protein manufacture can resume. We are back on track without lasting effects.

"The clever way we accomplish this is as follows. Some of our inventive bacterial colleagues, when present in the tissues of a human host, have found a way to synthesize nucleotides that doesn't require them to manufacture folic acid. They simply steal preformed folates, or one of the nucleic acid bases such as thymine or thymidine, from their human host. Imagine a sulfonamide roadblock along the folic acid synthesis highway. Off to the side of the main synthesis pathway, on the frontage road, trucks may be observed bypassing the roadblock, carrying thymidine or folate—actually stolen from the host tissues—to the protein-synthesizing factory at the far end.

"However, humans have come up with a clever wrinkle by adding to their antibiotic cocktail trimethoprim, a compound that also blocks nucleotide synthesis but at a position further down the pathway than that utilized by the sulfonamides. This double substitution thwarts our bacterial friends, who cleverly devise alternate pathways around the sulfonamide barrier only to be unexpectedly blocked later on by trimethoprim. We must go back to the drawing board to find an alternate pathway around the trimethoprim block as well.

"By the way, you should know that we have gained some satisfaction by pulling one sneaky trick in this process. As mentioned earlier, humans have invented the disk diffusion susceptibility test, in which, if you remember, antibiotic-impregnated filter paper disks are placed on an inoculated plate to determine

if we can grow at the margin. If none of our colonies appear in the area around the disk, humans consider us vulnerable, or what they call 'susceptible.' Presumably if we cannot grow around the antibiotic disk on a culture plate, we de facto will not be able to grow in infected tissue when that antibiotic is used therapeutically. Aha—not necessarily so! Picture what happens in the culture dish and how we pull off our trick. Thymidine is not present in the culture medium on which we are growing. We have no alternative way to synthesize DNA, and thus cannot grow around the sulfonamide disk; we appear to be susceptible. This fools the humans into believing that their sulfa antibiotic can be used therapeutically to stop our growth. But in vivo, in our tissue milieu, we use those truckloads of stolen pre-formed folate or thymine to sidestep the sulfonamide mole-cules that are trying to block our way. Thus, we are in vitro sus-ceptible, but in vivo resistant. Humans soon caught on to our ploy, but it was worth the emission of a few metabolic vapors of laughter in the interim."

Closing Comments: *Bacillus* JK

> The problem humans have with germs is that they work by rules that humans find hard to deal with, rules so different that before Pasteur no one knew what they were. Germs are quick; humans are slow. Germs have no thought for the future; humans plan. Germs have no technologies; humans are consummate users of tools. Most important, germs never give up. Humans do so all too readily.
>
> —Editorial, *The Economist*, May 20, 1995

As *Enterococcus faecium* concluded, *Bacillus* JK moved back up to the podium. "Thank you for that interesting account," it replied. "As parting words I wish to leave you all with a sense of considerable optimism concerning our ability not only to keep up with, but to stay ahead of, all human endeavors to find the 'final solution.' Although billions of our mates will continue to be lost at the hands of humans who persist in inventing new drugs, the last cell will never be sacrificed; a surviving resistant remnant will always remain. Even when our replication times are slowed down, the doubling numbers of these newly resistant remnants will be sufficient to suddenly and unexpectedly pop up in a new arena as a new 'emerging agent' of infection.

"As is true with my clan, many of us have developed more than one defense against antibiotics. One cell may have alterations in porin or ribosomal binding sites, produce beta-lactamases, and have efflux pumps operating, all at the same time. Very few antibiotics, even when more than one are used in combination, can cover this breadth of resistance. And through our conjugational telephone lines, we can always transmit some of our genes to fellow prokaryotes, or pick up some stray resistance-carrying plasmid or transposon along the way.

"We can also remain confident that humans will continue to lapse in carrying out certain preventive measures. Ferry services through human neglect will continue to allow us to jump from one human to another. Ineffective use of disinfectants will continue, when with applications at critical times humans could have literally removed us from the environment. The overutilization of antibiotics will not be tempered, as ill patients will demand such treatment when it may not be needed, and doctors will over- rather than undertreat in order to avoid the plague of attorneys. Such indiscriminate use of antibiotics gives us repeated opportunities to marshal new defenses, often never before used, repeatedly befuddling our human friends.

"Therefore, be of good cheer, my fellow prokaryotes. All remains well; our clans will survive. Thank you all for your attendance and the interesting contributions made by so many of you."

Final General Assembly:

The Renaming of *Homo sapiens*

There is no such thing as "official sanctions" by any body; since the science and understanding is continually evolving, it would seem undesirable. Names are now validated to the extent that the requirements of the Code are met and assessment that the science is good. Even then, what lasts is determined by general acceptance ... What is left to us these days, as we experience the ferment of ideas concerning the melding of phenotypic and molecular approaches to description, is to see that the education of microbiologists includes effective exposure to the nature and problems of bacterial taxonomy.

—R. G. E. Murray, *ASM News*, December, 1998

Thermotoga maritima

Chairbug Escherichia coli *called to order the final assembly: the last and perhaps most animated of the sessions. Attendance was almost perfect. Anticipation of the report of the Taxonomy Committee ran* high, to finally learn of their recommendations for a new name for Homo sapiens. *Although many names had been submitted for committee deliberation, it was decided by the rules subcommittee to limit the final list to only 10. Voting was planned to take place through several rounds, if necessary, until one name became the choice of over 50% of those assembled. Each of the 10 selected names would be entered for consideration by one or a small group of bacteria, which would briefly state the reasons for their choice. Genie Transposon waited eagerly at the side of the podium to post each suggested name as it was nominated.*

Before beginning the presentations of specific names, Escherichia coli *found it advisable to share certain of the committee deliberations that touched on broader considerations and implications.*

"Fellow prokaryotes," effervesced *Escherichia coli*. "It should be recognized here that the Taxonomy Committee expended considerable logarithmic metabolic efforts to sift through the many name change entries. Each name placed in nomination was given due consideration, and the selection of the final 10 was based primarily on what the committee members felt would have the greatest chance for final acceptance.

"It may be of interest to you to know exactly how, in the elevated circles of human deliberations, new bacterial names are finally determined. There is no 'official' classification of bacteria; that is, there is no designated human committee in perpetuity that sits on a high throne to stamp final approval on a name. Rather, a worker, after an appropriate period of study of the attributes and characteristics of the new species that he or she has discovered, publishes a proposed new name in one of a number of books or journals. If this new epithet meets all of the rules and regulations of nomenclature and taxonomy, and is not challenged by a peer or group of peers for a period of time, that name is eventually accepted. The importance of this explanation is that we prokaryotes will be publishing in this book our final name selection for *Homo sapiens*, which follows the prescribed protocol. Even if our proposal is never officially accepted in human circles, there will be at least a 'common law' adoption among ourselves.

"In this regard, I must challenge one statement made in a manual widely circulated among the human bacteriology community:

> It is important to emphasize something that has been mentioned often before; namely, that bacterial classifications are devised by microbiologists, not for the entities being classified. Bacteria show little interest in the matter of their classification. For the systemist, this is a very sobering thought!
>
> —James T. Staley and Noel R. Krieg, *Bergey's Manual of Systematic Bacteriology*, 1984

"A very sobering thought? Nothing could be further from the truth. Bacteria have great interest, if not in taxonomy as a whole, at least in the manner in which their names have been bandied about and debased through the decades. Judging from the rhetoric from the members of this assembly, it seems strong feelings and resentments persist over the capricious and insensitive manner in which humans have juggled the names of so many of our cohorts. In fact, the counterinvective from a few of our members was so strong that certain submitted names had to be summarily rejected simply because the chances for final acceptance were close to zero. This does not minimize, however, the feelings behind the clamor, nor does it necessarily mean that those suggestions would not be welcomed by many of the collective prokaryotes.

"In all fairness, I would say that the majority of the names applied to our bacterial friends have been reasonably well received. The majority of the designations given to the thousands of bacterial species that have been discovered and formally named to date are not really objectionable. Many names highlight one or more of our positive attributes, extol our ability to synthesize certain products, recognize the attractive appearance or distinctive aroma characteristic of so many of us, or focus on our ability to destroy or degrade certain products that may or may not be beneficial to humans and animals. For example, many of our families have names related to colors—*rubrum, roseum, flava, flavum* or *flavescens, luteum, citreum, niger,* and the like. What name could be more poetic than *Chromobacterium violaceum*?

"Many of our assigned names recognize our association with a variety of animals. This is somewhat of a two-edged sword for humans, as certain of our bacterial kindred, who live at peace in or on animals, become belligerent when inadvertently gaining access to humans. Such an animal-associated infection is known as a 'zoonosis.' But bacterial names that include designations such as *canis, equi, bovis, cati, ovis, suis, melitensis,* and *pullorum* carry for us appealing domestic or pastoral associations with dogs, horses, cows, cats, sheep, pigs, goats, and chickens, respectively. We recognize that certain of our clan, *Streptococcus equi, Mycobacterium bovis, Brucella suis,* and *Brucella melitensis,* to mention a few— pleasant as their names may sound—are nonetheless furiously combatted by human infectious disease specialists.

"Most of our prokaryote friends also are satisfied with names that indicate a particular talent each displays for performing certain tasks. Names that carry a suffix of *phaga* ('to eat'), *vorans* (the act of 'devouring'), or *faciens* ('doing, making, or causing to be made') generally are accepted by their recipients as individual insignias, much as a medieval warrior might carry a family coat of arms or an American Indian might adopt a name signifying a certain attribute (Singing Woman or Runs Like a Deer, for example). Examples of prokaryote names that recognize positive attributes include *Capnocytophaga,* particularly adept at utilizing ('eating') carbon dioxide (*capno* = 'smoke'; thus, 'cells that eat smoke'); *Serratia liquefaciens*, with its talent for liquefying gelatin; and, of course, *Pseudomonas acidovorans*, which is highly capable of utilizing ('devouring') all kinds of organic acids for its internal metabolism. Also, the prefix 'de' usually is favorably received. For example, *Alcaligenes denitrificans* and *Pseudomonas denitrificans* recognize talents for extracting nitrogen from amino acids and other nitrogen-containing compounds.

"*Azotobacter lacticogenes* was particularly pleased that both its nitrogen-fixing capabilities (*azo* = 'nitrogen'), so important for the nutrition and well-being of legumes and other food-bearing plants, and its ability to generate lactic acid (*lacti*, plus *genes*, or 'to generate') were formally recognized. However, its elation was short-lived (only 30 years in human time). A group of scientists and taxonomists joined forces and tagged our colleague with the new genus name *Beijerinckia*. Well, this honors Martinus Beijerinck, a highly respected microbiologist, but what does *Beijerinckia* have to do with their important talent of nitrogen fixation? It hardly seems fair.

"*Clostridium sporogenes* is content that everyone knows it has a unique capability of producing spores, and *Enterobacter aerogenes* is recognized for its ability to produce massive quantities of gas (you may remember that its original name was *Aerobacter*— thus, the double name *Aerobacter aerogenes* really emphasizes the wind power of this colleague). Several additional examples could also be cited. I mention these here only to illustrate that not all human-appointed bacterial names have negative or displeasing connotations.

"On the other side of the ledger, however, any name connoting a deficiency in the performance of duties or the inability to carry out certain functions is considered as a demeaning yoke. *Moraxella nonliquefaciens* and *Pseudomonas nonliquefaciens* are quite displeased at being singled out among their clans as being incapable of liquefying gelatin; similarly, *Sphingomonas* (formerly *Pseudomonas*) *paucimobilis* and *Edwardsiella tarda* are unhappy for being chided for their introverted propensity to ponder or to take things slowly.

"*Edwardsiella* is displeased with its first name as well; the association with P. R. Edwards, as famous and well liked as he was among American bacteriologists, still represents intrusion of a human name. Alas, names bearing '*ellas*' abound. This suffix *ella*, indicating 'little one resembling,' in and of itself is not offensive. It is the virtually universal association with a human that we find objectionable. *Salmonella* means 'a little one associated with Salmon,' that is, D. E. Salmon, an American bacteriologist. Similarly, other '*ellas*' are connected to historical bacteriologists from various countries—*Shigella* (K. Shiga, Japanese), *Klebsiella* (Edwin Klebs, German), *Bordetella* (Jules Bordet, French), and *Francisella* (Edward Francis, American). Our friends belonging to the *Legionella* clan are a bit less resentful, as their name is affiliated with a companionable group of fellows, the Legionnaires. But we object to being labeled with names of modern humans—respected though they may be—whose existence we predate by millions of years."

Escherichia coli looked up to see that flagella were flying and vapors were rising from those who could not move. It perceived that its discourse on taxonomy had strained the tolerance of an audience bent on getting to the main business at hand. It immediately raised several flagella to placate the gathering.

"Fellow bacteria," it announced, "your patience for a brief further explanation. You should know that the Human Taxonomy Committee decided against any attempt to change the genus name *Homo*. *Homo* means 'man,' and this cannot be altered. Therefore, only suggestions for a change in the species name were accepted. The name *sapiens* means 'wise' or 'sage.' Although it was agreed among the committee members that humans possess consciousness and have abilities for high-level thinking and reason, considerable discussion ensued as to whether modern man is truly 'wise' (or has ever been).

"The name *Homo faber* ('skillful man') has already been proposed in the human medical literature. In some philosophical circles, man is considered the 'maker and creator.' This race of *Homo* has charged itself with bringing about the moral transformation both of itself and of material things. Even this definition seems to fall short of modern humanity, within whom morality, honor, and honesty sometimes seem to be in decline judging from their war-like interactions, their disregard for family and human rights, and the modes of entertainment in which they indulge. The 'skills' of modern man, on the one hand, can be witnessed in amazing technological advances; yet, he seems to be more lost and uprooted than ever before.

"The name *Homo erectus* was also used at one time, in reference to the erect mode of ambulation characteristic of humans. However, this name has already been used to designate a now extinct *Homo*-like precursor that lived as long ago as 1 million years, but vanished around 200,000 years ago. According to paleontologists, *Homo erectus* had a smaller brain, only about two-thirds the size of that of *Homo sapiens*. The skull bones of *Homo erectus* were quite thick, the forehead receded sharply, the eyebrows protruded, and the nose, jaws, and palatine cavity were wide. It is still not known whether this species became extinct, for reasons not discovered, or transformed into an early *Homo sapiens*. Indications are that *Homo erectus* inhabited caves and mastered fire. *Homo neanderthalensis* and *Homo transvaalensis* (a new name for the apelike hominid *Australopithecus*) are earlier species names.

"I mention this brief history only to indicate that different species names for *Homo* have been introduced from time to time in the past, in keeping with evolution. Whether the recent evolution of *Homo sapiens* is sufficient to warrant an official name change remains to be seen. In the interim, we prokaryotes desire to offer our own suggestions.

"We will now proceed to the introduction of each of the 10 alternative names for *Homo sapiens* that are to be placed in nomination. At the end, we will ask each of you in the assembly to cast one vote.

"*Moraxella lacunata*, please commence the nomination process." As *Moraxella* rose, Genie Transposon leaped up to write this first recommendation on a scrap of paper on the kiosk.

Nomination #1: Submitted by *Moraxella lacunata*

"Thank you for the introduction, *Escherichia coli*. On behalf of myself and my colleague *Bacillus lacunatus*, we propose the new name *Homo bilacunatus*.

"As you may remember, the name *lacunatus* was applied to myself and my *Bacillus* colleague because we eat out the medium from under our colonies, producing what appears to humans as a pit or 'lacuna.' Our friend *Eikenella* (another of those '*ellas*') also shares this agar-eating capacity, but humans added the second name '*corrodens*' rather than '*lacunata*,' for reasons not immediately clear to us. Our reason for proposing this new name does not reflect any vindictive antihuman rebuttal; in fact, we have grown to accept this appellation cheerfully in recognition of one of our prized attributes.

"We reason that the accepted new name must apply to all humans. The human term 'lacuna' also refers to a 'hole' or 'opening.' As we observe humans, they appear from our vantage point to have a hole at both ends. In fact, many of our bacterial colleagues occupy one or both of those openings and therefore can verify that our observations are correct. For the most part, one opening seems to be related to yelping, talking, and imbibing; the other, with elimination. In some diseases, these processes seem to be reversed, which at times is a bit confusing to us. These particular inconsistencies, however, do not alter the observation that two openings well represent the norm. Therefore, '*bi* [two]-*lacunatus*' appears to be a reasonable designation for this universally shared anatomic feature, and we respectfully cast one vote for *Homo bilacunatus*. Thank you for this opportunity."

"Thank you, *Moraxella lacunata*. Your entry has been entered onto the master slate of candidates."

Looking over the throng, *Escherichia coli* saw that *Flavobacterium multivorum* was reflecting its yellow coat to be recognized.

"My apologies, colleague *Escherichia*," called out *Flavobacterium multivorum*, seemingly a bit reluctant to speak. "I know that we were not scheduled until later to place our selected name in nomination; however, the fact that *Moraxella lacunata* has put before this assembly a name describing a physical quality of humans, I thought it would be most in keeping to present ours at this time. Therefore, I petition you and this assembly, as a personal favor, to alter the order of scheduled presentations."

"As we have allotted sufficient time for this session," answered *Escherichia coli*, glancing at the list of names to be presented, "I see no reason why you should not go next, particularly in view of the name you are about to submit. Please continue, *Flavobacterium multivorum*." Hastily Genie Transposon scribbled the next name on the kiosk.

Nomination #2: Submitted by *Flavobacterium multivorum*

"Thank you. I also speak in behalf of colleagues who share our after-name—*Pseudomonas multivorans* and *Desulfococcus multivorans*." The former waved its head feather in recognition; the latter sent a few sulfur-rich bubbles to the surface from where it existed deep in the anaerobic mud bottom of Prokaryote Pond. "Actually," continued *Flavobacterium multivorum*, "we are quite pleased with our names. *Multivorans* or *multivorum*, you see, means 'the ability to utilize many carbohydrates,' a designation that we accept as a token of good fortune, as humans do not often so commemorate prokaryotes based on their innate talents. We therefore would like to reciprocate in kind, and are offering the name *Homo multivocalis* as our nomination.

"This name extension, of course, is in recognition of the many sounds that humans can emit. No other member of the animal kingdom, and certainly none of us microbes, has the capability to produce such a variety of sounds. Many animals make sounds, some quite beautiful. But for any family or group, the sound is only of a single pitch or sonority; for example, the bleating of a lamb, the whinny of a horse, the squeal of a pig, and the horn of a sea lion are sounds distinctive for each. Songbirds have a chirp or short medley, rarely more than three or four notes in sequence, and always the same for each species. Parrots that have been taught by humans to 'speak' perhaps come closest to this unique human trait. But even their 'speech' is purely imitative and limited to those few phrases learned from humans.

"Humans must be applauded for the several sets of genes they have accumulated through the ages to control their vocal cords, allowing an infinite variety of sounds. The soft lullaby hummed by a mother to her child, the resonating intonations in hundreds of languages by talented orators, the rapid staccato barking of an auctioneer, the sublime high notes of a lyric soprano, and the deep, sonorous reverberations from the throat of a bass baritone are unparalleled in any other single species in nature. Therefore, in recognition of this unique human quality, and in appreciation for their recognition of our talent, we reciprocate with the proposed name *Homo multivocalis*."

"Very poetically spoken," answered *Escherichia coli*. "Your depth of expression will make a favorable impression." It looked down at its list, and its flagella quivered in a brief fibrillation of surprise.

"Fellow bacteria," it said. "I myself submitted a name in nomination, and note here that I am scheduled next. That I should follow this most moving tribute by *Flavobacterium multivorum* is logical. Indeed, the name my clan has selected also recognizes a human trait that at times is quite beautiful. However, there are some differences in the manner in which the selection of the name we place in nomination came about. We do not exactly feel the same kindness toward humans as expressed by *Flavobacterium multivorum*." It waved a flagellum at Genie Transposon, who obediently wrote up the next name.

Nomination #3: Submitted by *Escherichia coli*

"My clan has been tagged with the name *Escherichia* in honor of our discoverer, Theodor Escherich, a little-known German physician working at the time of Robert Koch. Probably the name's coiner was a student who had a debt to pay a mentor. We would have much preferred a name such as *Bacillus flagellatus*, in recognition of the beautiful flagella that surround our coats, or perhaps *Bacillus lactofermentans*, highlighting our ability to ferment lactose, a task not accomplished by all prokaryotes. Though we do not enjoy being named for a fairly obscure human, we at least recognize the rationale behind the choice. But *coli*! No! We could certainly do better than being assigned a name that permanently associates us with the lower bowel segment of humans and animals. We occupy a wide variety of other niches that certainly could have been recognized. With the exceptions of *Campylobacter coli* and *Vibrio coli*, who also are not pleased with their second names, many other bacteria also reside in the lower bowel, but they are not called *coli*!

"We are not spiteful, however, and our name choice is based purely on physiological reality. We wish to place in nomination the name *Homo heterotrichoides*, in recognition of the fine and often profuse hair that covers most parts of their bodies. This entry perhaps reflects that we would have preferred a name such as *Escherichia trichoides*, in recognition of the fine, hairlike pili that cover our outer coat. Although some humans have no hair on the top of their heads, others possess thick dark, blond, or ruddy tresses that can even flow over the shoulders and back. Granted, not all humans have abundant hair. That is why we used the prefix '*hetero*': to indicate this difference. We initially thought of the name *Homo cephalocrina*, which literally means 'humans with hair on the head.' But this name would exclude all bald subjects, probably precluding its acceptance by the committee.

"My colleague *Gardnerella vaginalis*, to some extent in retaliation for the impolite commemoration of its site of occupancy, suggested the name *Homo pubiocrinosa*, rightfully arguing that virtually all adult humans have hair surrounding the pubic bone. Although our clan rejected that nomination on the basis that it would probably offend the Taxonomy Committee, let alone the human race, I nonetheless promised our friend *Gardnerella vaginalis* that we would at least pass on its sense of insult to the assembly. Therefore, from the ranks of the *Escherichias*, please record *Homo heterotrichoides* on the slate for vote by this assembly.

"Now, let us see. Who is next? You? Is that our good friend *Streptococcus agalactiae* that I see in the distance? Yes. Very well, the attention of the assembly is yours."

Nomination #4: Submitted by *Streptococcus agalactiae*

"Thank you, *Escherichia*. I shall be brief. I represent other colleagues whose names are preceded by the letter 'a,' meaning that there is something we cannot accomplish. For example, *Citrobacter amalonyticus* received its name because it cannot hydrolyze malonate, and *Achromobacter xylosoxidans* because it could not produce colonies with pigment. We feel that *Achromobacter* should be commended, not condemned, for not flashing an unsightly or garishly pigmented coat. The name *Acinetobacter anitratus*, a former unofficial designation for our eminent keynote speaker *Acinetobacter baumannii*—now fortunately discarded—carried a double whammy: couldn't move and couldn't reduce nitrates. But if you don't have the genes, you don't have the genes. That's all it amounts to.

"The story of my species name," continued *Streptococcus agalactiae*, "goes back to former days when we commonly infected the udders of cows and goats, rendering them unable to produce milk (*agalactiae*). Somehow this name has stuck with us through the years, although we would prefer something more positive, such as *Streptococcus campiae* in recognition of our capability of producing the CAMP factor, a unique substance that amplifies the beta-hemolytic reaction when we are grown on blood agar.

"We have attempted to rise above our sense of affront, but nonetheless feel a strong urge to reply in kind. Rather than striking back in an aggressive manner, as many humans do to express their grudges, we will express our discontent by highlighting, in turn, a defect that all humans have. Although *Escherichia coli*'s coinage of this name earlier was intended facetiously, we think it was a pretty good idea.

"Observing the graceful creatures of nature that take to the air, we note that the ability to fly is one feature that all humans lack. Thus, in recognition of this defect, we propose in all seriousness *Escherichia coli*'s suggestion of *Homo aflugialis*, or lacking wings. In truth, we decided on this entry only after long debate. *Citrobacter amalonyticus*, observing that humans, in contrast to most other animals, do not possess a tail, suggested the name *Homo acaudiforme*. After some discussion, however, we realized that the coccyx bone of humans represents the rudiments of an ancient tail, one that evolution has gradually hidden but is nonetheless still there. Therefore, '*acaudiforme*' is not technically correct. Thus, after due deliberation, we elect to place the name *Homo aflugialis* on the final list of entries."

"Thank you, *Streptococcus agalactiae*," replied *Escherichia coli*. "I suppose I could 'butt-out' and say, 'that will never fly'"—more gaseous groans arose from the audience—"but, upon second reflection, the name you have chosen flows well off the tongue and does point to a universally missing feature of the human race. Thus, we enthusiastically receive your entry.

"*Erwinia*? *Erwinia herbicola*, are you ready to come forth? You are listed next in line. Oh yes, there you are. The assembly is now open to receive your nomination."

Nomination #5: Submitted by *Erwinia herbicola*

"Thank you, *Escherichia coli*. Actually, I speak in behalf of a number of my cohorts who also carry '*cola*' in their after-names. *Cola*, you see, means 'to be associated with' or 'to dwell on or within.' Thus our name, *herbicola*, means 'grass or herb dweller.' In that same vein, *Spiroplasma floricola* ('dweller of plants') and *Pseudomonas hibiscicola* ('dweller of hibiscus,' if you really want to get specific) also have vegetation associations. Then, of course, there are all the dwelling places within the human body, presumably where we either lie in commensal peace or cause infections. Examples include *Bacteroides denticola* (in tooth crevices), *Vibrio costicola* ('rib dweller'), and the insultingly graphic *Herrelea vaginicola* (which, we are pleased to say, has been removed from the lexicon!).

"A few of the more aggrieved among us even tried to come up with a name that would reflect humans living in the sewer, gutter, or drain—*Homo piscinus* or *Homo pissier* were suggested, on the basis of the human taste for debauchery—but clearer heads realized that such vulgarity would never pass the nominating committee. *Pseudomonas gelidicola*, amused by the possibilities of this idea, suggested the name *Homo cocacola*. This, of course, was in jest; although, judging from what we observe in certain of the cartel circles, 'dwellers in coca' may not be that far from the truth, the name is meant to suggest the quantities of cola soft drinks that humans ingest daily worldwide. As one must be careful in human circles not to express a bias toward one commercial company, *pepsicoli* should at least be mentioned, loosely translated into 'aiding in digestion' or something like that.

"To bring this back to the main discussion, however, we propose the name *Homo terracola*; that is, 'dwellers of the earth.' We considered other alternatives, including *Homo terregens* and *Homo terregenes*, both of which also suggest a global dwelling; however, the suffix '*gens*' or '*genes*' gives the false impression that humans also created the world (which perhaps is not too remote from the assumptions of some of them). In keeping with devising a name that pertains collectively to the human race, 'dwellers of the earth' certainly fits. We know of no humans living in domains other than planet Earth. We therefore proudly place the name *Homo terracola* in nomination."

Nomination #6: Submitted by *Bacillus licheniformis*

"Hallelujah, amen!" blurted out *Bacillus licheniformis*, a crusty old soul who made no pretense of applying for formal recognition from Chairbug *Escherichia coli*. "Yes sir. 'From dust to dust,' that is the universal plight of humans," continued *Bacillus licheniformis*. "On behalf of my fellow *Bacillus* clan, and speaking also for our close cousins, *Clostridium* species, who differ only by preferring to live in the deeper stale air regions of soil and mud, we second the name *Homo terracola* submitted by *Erwinia herbicola*. You see, we the bacilli and the clostridia have been living in the soil for millennia, even over paleontological eras, lasting through all sorts of adversities because we can retract into our bomb-shelter–like protectors called 'spores.' If there is anything eternal in our natural environment, it could very well be spores.

"We would also like to place an alternate name into nomination, however: *Homo terracyclicus*. Actually, I must admit that it was *Streptomyces griseus* who, picking up on the 'dust to dust' theme, came up with this suggestion. Contributing this name is possibly one of the few good turns that *Streptomyces griseus* has done for its prokaryote brethren. You see, *Streptomyces griseus* has not been among our favorite neighbors. First of all, it just plain stinks, making it difficult for us to live in its neighborhood. Humans also have noted their odor, which they compare to a 'musty basement.' Well, why do they think basements are musty? Simply because that is one primary dwelling place for the *Streptomyces* clan and some other closely related associates, such as *Nocardia*, whose aroma we have already noted.

"The *Streptomyces* stench we perhaps in time could learn to tolerate. But the coup de grâce is their insistence on producing substances that either maim or outright kill many of us who trespass too far into their domain. 'No Trespassing' signs abound! In fact, humans have picked up on this malicious trait, extracting from streptomycetes certain substances that have since become a significant constituent of their antibiotic armamentarium. Certain of us, able to disregard the streptomycete odor, have inherited mechanisms to thwart the onslaught of these manufactured antibiotics as well, thereby allowing us to establish a foothold in various sites of infection. Thus, although we still consider *Streptomyces griseus* an offensive character, the benefits we have derived from learning to deal with its presence allow us to overlook its personal flaws and recognize its brilliance in providing our nomination, *Homo terracyclicus*, a suitable scientific representation of 'dust to dust.' "

"Thank you, *Bacillus licheniformis*," replied *Escherichia coli*, "even though your manner is a bit disruptive. Nonetheless, we place your suggested new name on the list as the sixth nominee; Genie Transposon, please post it with the others. I surmise this voting assembly will find it a bit of a challenge to choose between two cleverly conceived names, *Homo terracola* and *Homo terracyclicus*, both of which characterize humans quite well. In passing, I might mention that humans also marvel at times on how innovative and creative inspirations can arise from individuals whose personalities are not socially appealing.

"Next, we will hear from *Proteus mirabilis*, the very name of which bespeaks wonder."

Nomination #7: Submitted by *Proteus mirabilis*

"What has amazed myself and others in our *Proteus* clan, as we have studied the names that humans have applied to bacteria, both past and present, is the virtual absence of any reference to the heroes of human mythology. As you know, we proudly carry the name of Proteus, a water-dwelling herdsman for Neptune, the great Roman god of the seas. The magical Proteus had the uncanny ability to assume different shapes. In one account, he first assumed the form of fire, then flood, then that of a wild beast in order to escape the beekeeper Aristaeus, who, incidentally, was only trying to get some advice on how to recover his lost bees. I suppose certain humans, looking at the ways we grow on the surface of agar plates—at times appearing as entire colonies with well-defined margins, at other times extending ourselves as waves across the surface, and at still other times as a thin veil or mist—assumed we had inherited from Proteus a power to assume these different forms.

"As for others of the prokaryotes, had we been asked, we might have suggested to humans that a name reflecting some form of Cupid, the god of love, or the beautiful maiden Psyche should have been used for *Neisseria gonorrhoeae*, the organism causing the sexually transmitted disease gonorrhea. These two characters, in the myth, were the personification of the human soul and the target of its amorous arrows. Their interactions being far more conducive to acquiring a 'venereal' disease, *Cupidoides gonorrhoeae* or *Psychococcus gonorrhoeae* might have been more appropriate. Instead, this organism is named for Herr Albert Neisser, who probably never came close to having the 'clap' in his life.

"And instead of *Pseudomonas*, in recognition of their anterior polar flagella often appearing like multiple snakes it might have been more appropriate to use some form of Medusa, the beautiful mythological maiden transformed into a monster with snakes for hair. *Medusamonas aeruginosa* would have been a far less demeaning name than *Pseudomonas* and would have caught on in human circles just as well. This, of course, can never be, as the name *Pseudomonas* was inscribed in the annals of taxonomy long before humans were able to see or even knew about flagella.

"Or how about *Icarusiella* instead of *Klebsiella*, for these bacteria that have flagella as fragile as the wings of foolhardy Icarus, who lost his life when he flew so near the sun that their wax moorings melted? The flagella of these bacteria may also be lost or become dysfunctional when the bacteria reach a certain high temperature. *Icarusiella pneumoniae* may not be as flowing and poetic to the human tongue as *Klebsiella pneumoniae*, but from our perspective would be far more reflective.

"With this in mind, we propose *Homo narcissiosus* as the new name for *Homo sapiens*. Narcissus, the beautiful son of Cephisus, saw his own reflection in a fountain and thought it to be the lovely presiding nymph of the place. He eventually pined away for love of this unattainable spirit. Humans are much like that. Reflecting on themselves in the mirror of the universe, they perpetuate a self-perception as the pinnacle of all existence, pining away at their own grandeur. Yes, *Homo narcissiosus* is an appropriate name for them. We debated two alternatives, *Homo egogenicus* and *Homo animoides*, both equally reflecting this vainglorious and egocentric attitude. But we protei decided that the name of a character from the distant world of fantasy would better express our intent."

"Thank you, *Proteus mirabilis*," answered *Escherichia coli*, "that is indeed an interesting suggestion. *Stenotrophomonas maltophilia*, I now ask you to come forth and place the next name in nomination."

Nomination #8: Submitted by *Stenotrophomonas maltophilia*

"Thank you, *Escherichia coli*, for providing this opportunity. As the evolution of our designation, *Stenotrophomonas maltophilia*, has already been presented before this assembly, I will only briefly review some of the details in reference to the name for *Homo sapiens* we are pleased to put in nomination.

"The Latin stem *steno* means 'close,' 'narrow,' or 'little.' The word *tropho* means 'to grow'; thus, the '*Steno-tropho*' part of our name indicates that we are narrow or limited in our ability to grow. The fact is, there are only a few environments in which we prokaryotes, in general, *cannot* grow; thus, humans used this very narrow gap in our personal physiology as a basis to coin our current name.

"Picking up on that theme, then, we have similarly considered in what trait humans may be deficient. In contrast to prokaryotes, there are very few ecosystems in which humans can thrive. Many of our prokaryotic kin can survive without oxygen, in fact were among the first of living matter in that form. Many others thrive in saline concentrations approaching 12% (humans cannot withstand salt environments much beyond 'physiological saline'—0.85%), or environments reeking in sulfur, iron, and other natural elements. Crack open a dense rock and there you will find bacteria;

in samples taken from the deepest clefts of the ocean, or miles down in the crust of the earth, where there is no light or oxygen and the barometric pressures would crush a human, we still will be found.

"We already appreciate the ability of *Helicobacter pylori* to withstand the extreme acid of the stomach mucosa. Other acidophiles have been added to animal feeds, where they work in the acid-rich rumen to help animals digest their foodstuffs more efficiently. Still others of our brethren survive and work in environments of very high pH or alkalinity, which often is of use when working in conjunction with detergents to help clean up various forms of pollution.

"In fact, those of our prokaryotic colleagues colloquially referred to as 'extremophiles' occupy virtually every natural nook or cranny that has been investigated by humans. It is estimated that for every microbe species currently known to humans, there are 1,000 more that exist in some niche never examined. Humans must realize that the biomass of all the prokaryotes on earth exceeds that of all plants and animals combined! Therefore, in view of our observation—speaking specifically in behalf of our *Stenotrophomonas* clan—that the niches inhabited by humans are far more limited than those occupied by us and our prokaryote colleagues, we suggest that the name *Homo sapiens* be changed to *Homo stenohabitans*; that is, 'having limited habitations.' *Flavimonas oryzihabitans*, the name of which mistakenly indicates a habitat limited to rice and grains (*oryzi*), has agreed to second this nomination. Thank you again for the opportunity to enter a name in nomination."

"Well done, *Stenotrophomonas maltophilia*," replied *Escherichia coli*. "Expounding on our wondrous biomass not only has boosted our sense of importance in the world in general, but should also send a message to humans that we are always to be contended with, no matter where they may find us. Let us see. We are getting near the end of our nomination process. *Caseobacter polymorphus*, I believe you have a name you wish to place in nomination?"

Nomination #9: Submitted by *Caseobacter polymorphus*

"Yes, thank you, *Escherichia coli*. I shall be brief. I speak in behalf of *Brevibacterium casei*, *Leuconostoc mesenteroides*, and *Pediococcus acidilactici*, all of which have been utilized by humans in the dairy, pickling, and meat industries. We are not mentioning this to complain, but rather, on the positive side, to point out how many of us are of help to humans and in the improvement of animal life in general. We do not feel that we have been exploited; rather, that we have been honored for certain metabolic activities that make many foods more palatable.

"We the *Caseobacter* clan, along with our colleague *Brevibacterium casei*, play an important role in seeing that cheeses ripen properly, providing a unique flavor. The flavors of Limburger and Meshanger cheeses, perhaps not as palatable for some as others, nevertheless reach their peak only because of the hard metabolic work in which each of us is engaged. In particular, *Brevibacterium casei*, as its name indicates (Latin *caseus*, 'cheese'), works to see that Cheddar and cottage cheeses properly curdle from raw milk. We should mention that humans actually discovered our good work and have slowly and patiently cultivated us to make the best of cheese products. And in all fairness, in this process, they have provided us with a rich environment in which to grow, much like putting a herd of cows into a self-propagating pasture. We are, in fact, pleased with our symbiotic relationship.

"Further, the metabolic prowess of our colleagues *Leuconostoc mesenteroides*, *Pediococcus acidilactici*, and their relatives was also discovered to add considerable flavor to the many pickled and fermented products in which they dwelled. Also, they produce substances, particularly when included in meat products, that contribute to preservation and storage. Again, it is a mutually beneficial arrangement: we thrive in our limited niche, and humans value and encourage us.

"As we scan the broader range of humankind, we see that this is true in many fields of human experience, with much progress through the centuries in the improvement of life. More concern is being expressed nowadays over human rights, animal rights, and the preservation of the environment than at any time in history. Granted, villains have always existed among humans, just as they have at any other level of nature: what about the destructive antics of our *Salmonella, Shigella, Vibrio, Yersinia*, and other clans? But in most spheres of nature, an improved quality of life is found thanks to the opportunism, profit motives, and ingenuity of humans. Therefore, we submit for consideration the name *Homo expediens*, in recognition of the cleverness with which they have harnessed us to improve certain segments of living, and in recognition of their overall drive toward the improvement of life and environment."

Suddenly there was an eruption of emotional metabolic products from one side of the audience. "What am I sensing?" interrupted *Streptococcus pyogenes*, without waiting to be acknowledged by the chairbug. "What are the signals that so vibrate the proteins in my receiver domain? What is causing every part of my being, from my cell wall outside to my ribosomes inside, to so shudder? *Caseobacter polymorphus* and its crew of supporters are attempting to paint an unrealistic picture of humans. Humans are bacterial killers, I tell you—bacterial killers! How many of my clan have been mass murdered by the widespread use of penicillin and other antibiotics? Is this not also true for others of our kinsbugs—*Haemophilus, Neisseria, Staphylococcus*, and others? How many millions of our microbial mates have been lost, almost to the point of annihilation in some circles? No, humans are killers of microbes, that is all there is to it!"

"Please restrain yourself, *Streptococcus pyogenes*," burped *Escherichia coli* with some irritation. "You are out of order. If anything, your sudden outburst will do no more than engender sympathy votes for *Caseobacter*'s nomination. We listened to you patiently in the human disease session while you presented your

The Other End of the Microscope

justification for causing so much human misery. Our committee members judged that *Caseobacter*'s nomination was appropriate, as it focuses on the positive human contributions to our symbiotic relationships. Nonetheless, we will now honor your scheduled time to place in nomination your suggested name. Please proceed."

Nomination #10: Submitted by *Streptococcus pyogenes*

"Your rebuke may be warranted," *Streptococcus pyogenes* acknowledged, still agitated. "However, none of us bacteria should go soft on *Homo sapiens*. Too many of our families and clans have been nearly eliminated at the hands of humans and their antibacterial campaigns. Yes, we accept the risk that our nomination may be disqualified for being too offensive, even though we all agree that humans have stuck innumerable offensive names on *us*! Nonetheless, we hereby appeal anew to the Taxonomy Committee to at least let our nomination reach the assembly for a vote. We call for strong support from *Staphylococcus hominis*, *Cardiobacterium hominis*, and *Mycoplasma hominis*, all of which have had the name *hominis*, that is, 'related to man,' foisted on them. How can humans attach a name *hominis* to one of our kin, and then turn right around and do everything in their power to scatter them to the winds?

"No, even risking its immediate elimination, we nevertheless feel strongly about having the right to put into nomination one of the following three names: *Homo microbevorare*, *Homo prokaryocida*, or *Homo bacteriolytica*. The name *microbevorare* relates that humans 'devour' microbes; *prokaryocida*, that they 'kill' us prokaryotes; and *bacteriolytica*, that they 'liquidate' bacteria. Although each of these names carries our message, we recognize that we can only submit one name. Therefore, we enter the name *Homo prokaryocida*, indicating that through the use of disinfectants, germicides, and antibiotics, humans are waging a broad and continuous war directed collectively against us prokaryotes. The following account condemns itself."

> At a very early stage man developed a concept that contagious disease was caused by invisible living things . . . when two living bodies are closely united and one of these two exercises a destructive action on a more or less extensive portion of the other then we can say that "antibiosis" exists . . . it might be possible to obtain suitable extracts to kill bacteria causing diseases in the body.
>
> —Peter Baldry, *The Battle against Bacteria: a Fresh Look*, 1965

"Your nomination has been recorded, *Streptococcus pyogenes*," answered *Escherichia coli*. "I believe that this concludes the nominations. The time has finally arrived for each organism in this assembly to cast its vote. Please consult the list of candidates on the twig kiosk, from which you must choose."

"Before casting the final vote from among these listed candidates, we will allow one comment in succession for each entry, either in support or in rebuttal. Please keep your arguments brief, as darkness is about to descend, signaling the end of this assembly. I would encourage those of you who have not yet contributed to this congress to make your views, and those of your immediate clan, known. Although not a hard-and-fast rule, it may be most prudent to minimize bias by having only those of you without any human designation in your name to respond. In the interest of time, you need not ask recognition from the podium; rather, follow one after the other with your comments. Let the process begin."

"*Bacterioides fragilis* reporting. Thank you. As representative of the anaerobic bacteria, and particularly with our main niche in the tail end of humans, I would like to speak to *Moraxella lacunata*'s entry of **Homo bilacunatus.** Similar to the argument given by *Streptobacillus moniliformis* that *Homo heterotrichoides* is not

Nomination #1: *Homo bilacunatus*
Nomination #2: *Homo multivocalis*
Nomination #3: *Homo heterotrichoides*
Nomination #4: *Homo aflugialis*
Nomination #5: *Homo terracola*
Nomination #6: *Homo terracyclicus*
Nomination #7: *Homo narcissiosus*
Nomination #8: *Homo stenohabitans*
Nomination #9: *Homo expediens*
Nomination #10: *Homo prokaryocida*

unique to the human race, I would argue that the term '*bilacunatus*' is not specific. All animals of the higher kingdoms, and even those belonging to lesser families and orders, also have oral and anal or cloacal 'ends,' anatomical parts that I would submit are neither structurally nor functionally distinctive. Therefore, I recommend a vote against this entry."

"*Lactobacillus acidophilus* would like to cast one vote in support of *Flavobacterium multivorum*'s entry of **Homo multivocalis.** We agree totally with the argument put forth by *Flavobacterium multivorum* that the human voice is unique among animate and inanimate nature. It possesses a diversity of sound and expression found nowhere else in the animal kingdom, at least in the realm of sound waves and acoustical effects. We urge all of you to consider carefully submitting your vote for *Homo multivocalis*, which also has a pleasant-sounding 'ring.' "

"*Streptobacillus moniliformis* here. Representative of the pleomorphic, filamentous bacteria, I have particular interest in *Escherichia coli*'s nomination of the name **Homo heterotrichoides.** Despite my innate desire to have a name that might reflect something 'hairy' or 'filamentous,' I would nonetheless recommend voting against this entry. Although *Escherichia coli*'s argument is interesting, many animals have pelts and furs that are far more diverse and comely than the hair grown by most humans. Therefore, I do not see this as a unique human trait and cannot give my support."

"Excuse my intrusion," inserted *Aeromonas hydrophila*. "I believe that, acknowledging the fact that many animals have plumes and furs far more appealing than the hair displayed by most *Homo sapiens*, we also cannot support this nomination. The ability to cool off by sweating, on the other hand, is an attribute more unique to *Homo sapiens*. Therefore, we request that *Escherichia coli* reconsider and place in nomination the name *Homo hydroderma*, a name far more universally in keeping with the human race."

"Sorry, *Aeromonas hydrophila*," replied *Escherichia coli*. "We have made the rule that there will be no further nominations accepted from the floor that have not passed consideration by the Taxonomy Committee. This is an issue that you should have brought to my attention before this session began. I regret, but we simply must keep our rules. *Abiotrophia defectiva*, you are now recognized for the next comment."

"Yes, thank you. As our assigned species name, *Abiotrophia defectiva*, also points to a defect—that is, that we do not always grow well unless given all the nutrients we need—we would like to speak in favor of *Streptococcus agalactiae*'s entry of **Homo aflugialis.** Although a name focusing on other human defects may have been selected, we are satisfied that *Streptococcus agalactiae* and its advisors scanned other possibilities and decided that the lack of wings points to one glaring omission.

One possible argument against this name, however, is the fact that humans do 'fly' and have invented 'wings,' even though these are not part of their immediate anatomy. Therefore, taken in a broader context, the name *aflugialis* may not totally reflect reality."

"*Comamonas terrigena* has the following comment. As our second name, 'from the land,' also is associated with earth, we would like to speak in favor of *Erwinia herbicola*'s suggestion of **Homo terracola.** Being firmly attached to earth is certainly a universal human trait. Perhaps it can be argued that other members of the animal kingdom are also 'of the earth' and thus such an association is not unique. Nonetheless, of all the entries, this name has the most appeal to us. We recommend a vote in favor of *Homo terracola.*"

"*Agrobacterium radiobacter* would like to follow up on and respond to the comments just made by *Comamonas terrigena*. As our first name indicates, we are also 'attached to the earth,' and therefore we would like to give support to *Comamonas*'s recommendation. However, since a comment has already been made on that entry, we would also recommend support for *Bacillus licheniformis*'s entry of **Homo terracyclicus.** Although animals certainly also follow a 'dust to dust' cycle, this trait in particular has been specifically related to humans, at least if biblical lore has any bearing on this process. The words from the Genesis story of humankind itself give strong support to this entry: 'You shall eat bread till you return to the ground, for out of it you were taken; you are dust, and to dust you shall return.' Thus, we cast one vote in favor of *Homo terracyclicus.*"

"*Obesumbacterium proteus* has the following comments. First of all, we are not happy that some human along the way has called us 'fat' (*obesum*). However, the attachment of '*proteus*' to our name obviously piques our interest in *Proteus mirabilis*'s entry of **Homo narcissiosus.** Deep within our ribosomes we would like to see a mythical figure so immortalized. However, on reflection, we consider that the activities of so many mythological figures have been recorded that to single one out would be a disservice to the rest. Secondly, mythological figures are closely related to and patterned after humans; thus, a name such as *Homo narcissiosus* in effect merely doubles two human names. Additionally, not all humans have narcissistic tendencies; therefore, this is not a universal trait. Therefore, reluctantly, we recommend a vote against this entry."

"*Mycobacterium thermoresistibile*, representing the acid-fast bacteria, would like to speak in favor of *Stenotrophomonas maltophilia*'s nomination of **Homo stenohabitans.** We might have preferred the name *stenothermohabitans*, indicating the very narrow temperature range within which humans can survive; however, we would agree with *Stenotrophomonas*'s comment that such a more restrictive name would rule out other places of habitations also hostile to humans. Therefore, we concur with the more broad and generic name *stenohabitans*, indicating that humans have a far worse time of it than the prokaryotes in finding just the right environment in which to flourish. And just to remind humans one more time that should any great universal ice age or worldwide burn occur, we cryophilic and thermophilic microbes will still be around to continue life into the next age."

"*Methylobacterium extorquens* is ready to report. As we also have been used in an expedient way by humans to extract (*extorque*) methane from a variety of sources and have been of assistance in cleaning up oil spills, we are drawn to *Caseobacter polymorphus*'s contribution of **Homo expediens.** We can understand *Streptococcus pyogenes*'s strong rebuttal, and perhaps can also agree that some of us, who have not caused humans any misery and who have been used creatively to improve the human condition, may have a pacifistic orientation. On the negative side of this name, not all humans are 'expedient'; therefore, this also may not meet the criteria for universality. However, the same argument can be made for *Homo sapiens*, as not all humans are 'wise.' "

"*Erysipelothrix rhusiopathiae* wishes to be recognized, as contributing the last testimony in this process. I basically represent myself, as few other prokaryotes cause the same effects on humans that I do. As my name indicates, I am a slender but not exactly hairlike ('*thrix*' is a bit misleading) bacillus that produces inflammatory and ulcerative lesions that result in reddening of any skin into which we penetrate. There is not time to get into the argument here that humans should know how to prevent direct contact with me and my close associates, thereby preventing disease. However, once discovered, we are bombarded with antibiotic agents to erase us from the scene. In this regard, we would agree with *Streptococcus pyogenes* that humans, from our perspective, are out to rid the world of bacteria, at least those of us whom they consider dangerous, and therefore will cast one vote in favor of the suggested name ***Homo prokaryocida.***"

At this point in the bacterial assembly, the proceedings went relatively rapidly. Soon after completion of the last testimonial, *Escherichia coli* called for the first round of votes. It was made clear that the name to be selected must receive more than 50% of the vote, and that successive rounds of voting would be held until that goal was achieved. As no one entry achieved a 50% majority after the first round, the five entries receiving the lowest number of votes were eliminated. Remaining for the second round of votes were *Homo multivocalis, Homo terracola, Homo terracyclicus, Homo stenohabitans,* and *Homo prokaryocida.*

Clusters of activity were observed before the second round of voting was held. After the second round, only two names received the majority of votes, although neither exceeded 50%. Dropping by the wayside were *Homo terracyclicus, Homo terracola,* and *Homo prokaryocida.* Therefore, the final vote was between *Homo multivocalis* and *Homo stenohabitans.*

Escherichia coli allowed a few more comments from the assembly in support of the final two candidates. After all final votes were tallied, *Homo multivocalis* received 52% of the vote, and *Homo stenohabitans,* 48%.

Escherichia coli then proclaimed, "Be it known, then, that ***Homo multivocalis*** shall be placed in the formal records as the prokaryotes' new name for *Homo sapiens*, as duly nominated and passed in this assembly. Secondly, be it decreed that formal notice shall be made public, following the time-honored practice of indicating proposed name changes through official reviews, by means of the proceedings of this congress as here recorded.

"I thank you all for your interest and attention, and your participation in this historic assembly. I bid you all good evening." With a vigorous waving of its flagella, that regally reflected the last remaining rays of the setting sun, *Escherichia coli* declared the close of the assembly.

Epilogue

... No epilogue, I pray you, for your play needs no excuse;
for when players are all dead, there need none be blamed.

—Theseus, in William Shakespeare,
A Midsummer Night's Dream

 Our narrator, *Thermotoga maritima*, offers the following analysis of the proceedings for its human readers.

"And so, good readers, the assembly ends. We prokaryotes have only a few remaining thoughts to leave with you our good *Homo multivocalis* friends, as now newly named. You may imagine that Prokaryote Pond will remain, along with its bacterial participants, long after this congress. Several thousand years will be required for many of these organisms to live out their normal lifespans, before the stationary growth phase finally brings them to an end. But a basic ecological principle dictates that any natural or human-made disruption in a given region, be it large or small, will ultimately return to its pre-injury state, given sufficient time. Thus, a few centuries hence, a passing life form observing Prokaryote Pond will distinguish no disruption in the surrounding terrain, and never realize that the once-in-an-eon event just described ever occurred.

"And so, what do I, *Thermotoga maritima*, consider has been accomplished by this unusual gathering just chronicled? I can only share my inner reflections, as specific goals and objectives were never spelled out at the onset, except for the single purpose of advancing a new name for *Homo sapiens*. If that was our mission, I suppose we can with some pride conclude that this goal has been accomplished. Of equal importance was the opportunity afforded to many bacterial clans to vent their inner agitations and, on the positive side, to relate—often quite eloquently—their perceptions of their importance in creation, including notably the potential role of ancient prokaryotic survivors as essential elements in the intracellular organelles of every eukaryotic being. If the evolutionists finally prove that mitochondria and cilia, which serve such integral and necessary functions in eukaryotic cells, indeed have been derived from some ancient bacterial primordium, the integral necessity of prokaryotes in the scheme of creation is only reemphasized.

"Recording the proceedings of this assembly may serve to benefit our human readers as well, at least to help them understand that bacteria have been in existence since the onset of measured time, and furthermore, that they have the genetic tools necessary to remain until the last pulse of existence. The public orations in this assembly by many of the speakers have iterated over and over again that the infective diseases afflicting humans by and large result from human failings, and certainly not from any premeditated aggression on the part of the bacteria. Any creature devoid of a nucleus has no innate volition, but can only play out the genetic cards it has been dealt. We can do no more than follow unfailingly the patterns set by our genes, whether they be to sustain the metabolism of the organism or to carry through the production of substances that may be of harm to humans.

"Prokaryotes and humans are in this world together and must reconcile their interactions, to their mutual benefit. Obviously it is of little use for a bacterial rogue to kill off an infected human host. Not only will the organism commit self-elimination, but those humans not succumbing to the lethal onslaught will live on, immune to any future invasion. The only possible solution is for the pathogen to domesticate its infected host, with the ultimate goal of perpetuating its clan and living on as an attenuated commensal or symbiont. Such domestication also is of import to the human host, as is evidenced by the ancient respiratory bacteria that have persisted in eukaryotic cells in the form of mitochondria. What if eons ago either the initial eukaryotic cell or the respiratory prokaryote had given up in the process?

"There is no illusion on our part that the new species name *Homo multivocalis* will ever be accepted in human circles, or that the other deliberations at Prokaryote Pond will even be seriously considered. Nonetheless, if the name *Homo sapiens* stands, humans must then assume the role of 'the wise' and view the world of the prokaryota differently than in the past.

"We acknowledge that there are rogues among our ranks that, either from a chance chromosomal mutation or through the acquisition of plasmid or transposon DNA, turn vicious and cause considerable human disease and misery. However, this must be balanced against the great benefits that humans derive from our omnipresence and hard work.

"First of all, we were among the first to be involved in the process of fermentation; in fact, we can be credited with 'inventing' fermentation. You must understand the definition: for the most part, fermentation is an essential process for the energy needs of all living matter, not just a mechanism for making intoxicating beverages. If we were not continuously at work taking nitrogen from the air and 'fixing' it in the soil and oceans, all organisms, including the foliage and plants on which humans depend for much of their well-being and survival, would weaken and ultimately disappear. If it were not for our involvement in photosynthesis, which is another process that our ancient ancestors 'invented,' the world oxygen stores, so vital for human existence, would also begin to wane.

"We inhabit many parts of the human anatomy, serving as benevolent soldiers to ward off attacks from foreign microbes, particularly the fungi but also including some of our own militant colleagues. Once we leave the scene, primarily when we are eliminated by the overuse of antibiotics, these resistant invaders come in hordes. In fact, it has been brought to our attention that certain inventive humans are actually culturing some of our colleagues, lactobacilli to be exact, putting them into bacterial suppositories for reintroduction into sites such as the vaginal canal to ward off the invasion of yeasts.

"We have been used to help clean up oil slicks that humans dump along pristine coastlines. We were the first, through the acquisition of orange and purple pigments, to introduce the carotenoids, which find their way in the form of vitamin A into a variety of nourishing foods such as carrots. All sorts of living creatures, from insects through animals, birds, and humans, use these carotenoids to make rhodopsin, the pigment necessary for visual acuity. In fact, rhodopsin—and therefore sight in general—is thought to have its precursor in the protein bacteriorhodopsin, which the archaeobacteria use to generate energy from light.

"It may be that we the prokaryotes will carry the torch on into the next million years of existence, long after humans, like the dinosaurs, have become extinct. Humans have it within their present power to annihilate themselves, either through the release of high-level radiation from nuclear attacks, or by eliminating the ozone layer by the release of pollutants at an alarming exponential rate. However, we prokaryotes have been surviving radiation effects for millions of years. In fact, it may be of interest to some of you that one of our colleagues, which humans have named *Micrococcus radiodurans* for obvious reasons, thrives in the radioactive waters used to cool nuclear reactants. Most discouraging to humans must be the fact that radiation actually stimulates the transfer of genes from one bacterial cell to another, thereby enhancing our ability to survive and share our characteristics, whether beneficial or aggressive, with many of our colleagues.

"Given sufficient time, bacterial pathogens through natural selection can significantly temper their virulence. In fact, we prokaryotes must self-servingly reduce our virulence in order to survive, since we too go down with the ship when an infected host is annihilated. Cases can be cited where highly virulent bacterial

The Other End of the Microscope

strains, capable of causing untold mortality and morbidity among humans, suddenly, after only a few thousand generations, become docile commensals. The loss of, or the inability to conjugate and pass on, a virulent replicating plasmid may be one explanation for this loss of pathogenicity. Or, the host may mount non-lethal defense mechanisms against a microbe, tempering its invasiveness or reducing the production of toxins or harmful enzymes, leading to a mutually healthy symbiosis.

"The chance that humans take in this arrangement is if there is a breakdown in any of their defense mechanisms—loss of immunity, decrease in white blood cell production, reduction in opsonins, and so forth. Unfortunately these altered states reflect more and more the human condition. Prolonged lifespans are accompanied by debilitation; long-term therapeutic regimens reduce the number of circulating scavenger cells; chronic illnesses produce malnutrition and wasting; and the compromise of immunity, either naturally or from the effects of disease or suppressive drugs, can be fatal. In each instance, humans will continue to face opportunistic invasion from microorganisms that, from the organisms' own viewpoint, are merely carrying out their natural functions.

"With this, my good reader, our allegory is ended. And as with all allegories, the reader must sift through what is fact and what is fiction. The bacteria telling their story cannot distinguish what may be real for humans. Humans, however, do know for the most part how microbes can directly and indirectly affect them, either for good or for ill, and how they may succeed or fail in combating their innocent intrusions. These facts are recorded in thousands of pages of medical literature. Microbes do not read books, and will continue to surprise by presenting themselves in forms never before encountered, to be further chronicled in story after story of 'new and emerging diseases.' It is to be hoped that, after observing this event taking place on the imaginary microscope stage called Prokaryote Pond, through a lens that reveals the perspective of the bacteria, perhaps humans will begin to look on the tiny specks in their own microscopes with a new understanding. These tiny specks, the 'gram blues' and the 'gram reds,' spherical or elongated in form, arranged in chains or in oriental letters or haphazardly with no sense of organization—these minute, complex creatures are an integral part of a larger domain, a realm that must be shared with all living matter in which DNA replication is the sustenance of life."

We came into the world like brother and brother;
And now let's go hand in hand, not one before another.

—Dromio of Ephesus, in William Shakespeare,
The Comedy of Errors

Further Reading

Aztruc, J. A. 1754. *A Treatise of Venereal Disease.* Innys, Richardson, Davis, Clarke, Manby and Cox, London, United Kingdom.

Balory, P. 1976. *The Battle Against Bacteria—a Fresh Look.* Cambridge University Press, Cambridge, United Kingdom.

Barns, S. M., C. F. Delwiche, J. D. Palmer, S. C. Dawson, K. L. Hershberger, and N. R. Pace. 1996. Phylogenetic perspective on microbial life in hydrothermal ecosystems, past and present. *CIBA Found. Symp.* **202:**24–39.

Buret, F. 1891 (vol. I), 1895 (vol. II). *Syphilis in Ancient and Prehistoric Times,* translated by A. H. Ohmann-Dumesnil. F. A. Davis Co., Philadelphia, Pa.

Dawkins, R. 1998. *Unweaving the Rainbow—Science, Delusion and the Appetite for Wonder.* First Mariner Books, Houghton Mifflin Co., New York, N.Y.

De Kruif, P. 1926. *Microbe Hunters.* Harcourt Brace Jovanovich, San Diego, Calif. (1954 edition).

Dennie, C. C. 1962. *A History of Syphilis.* Charles C Thomas, Springfield, Ill.

Doetsch, R. N. 1960. *Microbiology History.* Rutgers University Press, New Brunswick, N.J.

Dubos, R. 1962. *The Unseen World.* Oxford University Press, Oxford, United Kingdom.

Gladwin, M., and B. Trattler. 1991. *Clinical Microbiology Made Ridiculously Simple,* 2nd ed. Medmaster, Miami, Fla.

Lorian, V. (ed.). 1996. *Antibiotics in Laboratory Medicine,* 4th ed. The Williams & Wilkins Co., Baltimore, Md.

Margulis, L., and D. Sagan. 1986. *Microcosmos—Four Billion Years of Microbial Evolution.* University of California Press, Berkeley, Calif.

Needham, C., M. Hoagland, K. McPherson, and B. Dodson. 2000. *Intimate Strangers: Unseen Life on Earth.* ASM Press, Washington, D.C.

Rosebury, T. 1973. *Microbes and Morals—the Strange Story of Venereal Disease.* Viking Press, New York, N.Y. (*See chapter 13, "The Famous and the Infamous."*)

Stanier, R. Y., E. A. Adelberg, and J. L. Ingraham. 1976. *The Microbial World.* Prentice Hall, Englewood Cliffs, N.J.

Thomas, S., and T. H. Granger. 1952. *Bacteria.* Blakiston, New York, N.Y.

Willcox, R. R. 1949. *Venereal Disease in the Bible. Br. J. Vener. Dis.* **25:**28–33.

Index